Acoustics
for You

Acoustics for You

by

James H. Prout

Associate Professor
of Engineering Research
Penn State University
University Park,
Pennsylvania

Gordon R. Bienvenue

Associate Professor
of Audiology
State University of
New York
College at New Paltz

KRIEGER PUBLISHING COMPANY
MALABAR, FLORIDA

Original Edition 1990
Second Printing 1991 with corrections

Printed and Published by
ROBERT E. KRIEGER PUBLISHING CO., INC.
KRIEGER DRIVE
MALABAR, FLORIDA 32950

Printed in the United States of America.

Library of Congress Cataloging-in-Publication Date

Prout, James H.
Acoustics for you / by James H. Prout and Gordon R. Bienvenue.
p. cm.
Bibliography: p.
Includes index.
ISBN 0-89464-328-2 (alk. paper)
1. Sound. 2. Acoustical engineering. I. Bienvenue, Gordon R.
II. Title.
QC225.15.B54 1989
620.2—dc19

88-7699
CIP

10 9 8 7 6 5 4 3

TABLE OF CONTENTS

PREFACE

This textbook was written to provide an introduction to the principles of acoustics for those students who have little or no background in engineering. The material is especially directed to students in the speech and hearing sciences, although those in other health-related fields and theater arts may also find the approach useful.

The authors wish to thank Dr. R. L. Kerlin for supplying the material on Whole-Body vibration. The Appendix on Behavior and Well-Being was adapted from Community Noise Fundamentals: A Training Manual by P. L. Michael, W. T. Achor, G. R. Bienvenue, D. M. DeJoy, R. L. Kerlin, A. H. Kohut, and J. H. Prout; a manual written under a grant from the Environmental Protection Agency.

Grateful appreciation is extended to Vic King and Ed McGarvey who executed the original art work and to M. Leanne Zindler for her time and talent in the production of this manuscript.

James H. Prout
Gordon R. Bienvenue

INTRODUCTION

What is Sound?

Sound is a result of oscillations in pressure, stress, particle displacement and particle velocity in an elastic medium. In air, the sound is usually described in terms of changes in pressure that alternate above and below atmospheric pressure. These pressure changes are produced when vibrating objects (sound sources) cause regions of high and low pressure that propagate away from the sources.

The characteristics of the sound depend upon the rate at which the sound source vibrates, the amplitude of the vibration and the characteristics of the conducting medium. A sound may have a single rate or frequency of pressure alterations; however, most sounds have many frequency components and each frequency component may have a different amplitude.

Acoustics is a science that covers the generation, reception, application, and effects of sound.

CHAPTER 1
MEASUREMENT AND THE LAWS OF MOTION

INTRODUCTION

The study of any new science is primarily a language learning task. In order to study acoustics, it will be necessary first to learn some basic components of the language of physics, especially that of the study of motion. This chapter will consider the way physicists describe certain motion phenomena that are essential to the study of sound. The terms used to designate the various aspects of these phenomena have very specific meanings. By developing definitions for these words, we will be examining the details of these occurrences. We will begin learning the language of physics by defining some of the words that physicists use.

Let us begin by considering what we mean by the science of physics. The word itself comes from the Greek word *physikos* which means nature. It referred to the study of those things found in nature; what we today refer to as the general term *science*. As man's knowledge of the physical world around him grew, it became too great a task for one man to know expertly all aspects of science. Thus, the natural philosophers (general scientists) of the Renaissance gradually gave way to the chemists, astronomers, biologists, geologists and other specialized scientists of the nineteenth and twentieth centuries. The name physics was retained for those aspects of science that had not branched off to become a specialty field. Thus, we find that modern day physics is concerned with motion, heat, light, magnetism, electricity, sound and the atomic structure of matter, among others. While at first glance this may seem a motley collection of topics for study, further examination shows a unifying theme among all these subjects. In all cases, these topics are concerned with matter and energy. Thus, the present day science of physics is a study of how various forms of energy can affect those things that we see in the world around us. Since this text is concerned solely with acoustics, this chapter will introduce only those concepts necessary for the study of sound. While heat, electricity and magnetism are related to certain aspects of acoustics, they are not necessary for the study of its fundamentals so they will be dealt with in later chapters. After a brief introduction to systems of measurement, the remainder of this unit will deal with the phenomena of motion, while a specific discussion of sound will be presented in Chapter 3.

MEASUREMENT

Measurement of a quantity involves comparing that quantity with some established standard quantity of the same type. For example, suppose we want to measure the height of a pole and suppose we choose a measuring instrument calibrated in units of one foot. When we compare the pole with our measuring instrument, we find that 6 one-foot units reach exactly to the top of the pole. We conclude that the pole is 6 feet high.

A one-inch standard of length would have shown that the pole is exactly 72 inches high. Using a metric standard of length, we discover that the pole is 1.83 meters or 183 centimeters high. Note that two pieces of information are necessary to describe the height of the pole: a *number* and a *unit*. The numbers 72, 6, and 183 by themselves have no meaning without the unit of measurement: inch, foot, or centimeter. The height of the pole must be described as: 72 inches, 6 feet, 183 centimeters, etc. In calculations, the units may be multiplied, divided, or cancelled. The same units or dimensions must appear on each side of a balanced equation. This is often a good way to check that an equation is valid. To convert from one system of units to another, a unity multiplier can be formed from the equivalent units. For example, since 1 inch = 2.54 centimeters, the height of the pole in the example above can be converted from one system of units to another by proper multiplication by unity factors. If we have measured the pole to be 6 feet in length, we can find the length in centimeters as follows:

$$6 \text{ ft} \times \frac{12 \text{ inches}}{1 \text{ foot}} \times \frac{2.54 \text{ cm}}{1 \text{ inch}} = 183 \text{ cm}$$

The fractions 12 inches/1 foot and 2.54 cm/1 inch are *unity fractions* since the denominator is the same amount as the numerator. Multiplication by unity will not change the value of the original measured quantity. The units can be cancelled when they appear both in the numerator and denominator until we are left with the desired unit; in this case, *centimeters*. After performing the numerical calculations, we find that the length, 6 feet, is the same as 183 centimeters. Since these conversion factors are unity, they may be used either side up to perform the necessary conversions.

SYSTEMS OF UNITS

All mechanical quantities can be expressed in terms of three fundamental quantities: mass, length and time. The units assigned to these three quantities are peculiar to each system of units. The English system of measurement employs the *foot* for the unit of length, the *pound* for the unit of mass and the *second* for the unit of time and is often referred to as the foot-pound-second system. Within this system, the unit of force is the *poundal* whose units are pound-foot/sec^2.

Note that in the example of the measurement of the pole above, the length of 6 feet is equivalent to 72 inches, which also equals 2 yards. In addition to changing the units when determining equivalent measures within the English system, the numbers also change. When the same pole is measured in centimeters, the length can be expressed as 183 centimeters, 1.83 meters, 1,830 millimeters or 0.00183 kilometer. This system of units is called the *metric system.* Whenever the unit is changed, it is necessary only to move the decimal point an appropriate number of places.

Although the metric system takes many forms, all of the forms are based on decimal arithmetic so that units of different sizes can be formed by multiplying or dividing a single base value by powers of 10. In other words, changes in units can be made simply by moving the decimal point. Quantities used today in the study of acoustics use the MKS system of metric units. The MKS system is based on the fundamental metric units of the *meter* as the unit of length, the *kilogram* as the unit of mass and the *second* as the unit of time. When the fundamental units of *ampere* (A) for electric current, the *kelvin* (K) for temperature, the *mole* (mol) for the unit of quantity (amount of substance) and the *candela* (cd) for the unit of luminous intensity, these units make up the International System of Units, abbreviated SI in all languages.

The advantage of using the SI system is that all derived units are also unity in value although they are given new names. For example, the unit of force which we will be using is the *newton* (N), defined as:

$$N = kg \cdot m/s^2 \, . \tag{1.1}$$

In words, one newton is the force applied to a mass of one kilogram which gives the mass an acceleration of one meter per second^2. Notice that the magnitude, unity, is preserved throughout the definition. The meaning of the newton as a unit of force will be easier to understand later in our discussion of motion.

Work is defined as a force acting over a distance and, in the SI system, is measured in *joules* which are defined as:

$$\mathrm{J} = \mathrm{N} \cdot \mathrm{m} \,. \tag{1.2}$$

In words, this expression shows that one joule of work is done when a force of one newton acts over a distance of one meter. Again, the magnitude of unity is preserved in this definition. The rate of doing work is called *power* and is given the unit name *watt* which is defined as:

$$\mathrm{W} = \mathrm{J/s} \,. \tag{1.3}$$

In words, one watt of power is used when one joule of work is performed in one second.

These are the units most often encountered today in the study of acoustics and are summarized in Table 1.1. Other special units of the International System may be found in basic physics texts. Before the acceptance of the SI system of units, other fundamental metric units were widely used. This earlier system is referred to as the CGS system and is based on the fundamental units of the *centimeter* (0.01 meter) the *gram* and the *second*. Using these units, the unit of force is the *dyne* which is defined as:

$$1 \text{ dyne} = 1 \text{ gm} \cdot \text{cm/s}^2 \,. \tag{1.4}$$

Stated in words, one dyne is the force which will give a mass of one gram an acceleration of one centimeter per second^2. In the CGS system, the unit of work is the *erg*, which is defined as:

$$1 \text{ erg} = 1 \text{ dyne} \cdot \text{cm} \,. \tag{1.5}$$

In words, one erg of work is performed when a force of one dyne acts through a distance of one centimeter. The unit of power is the *watt* which is defined as:

$$1 \text{ watt} = 10{,}000{,}000 \text{ ergs/sec} \tag{1.6}$$

and here, as you can see, the system begins to get out of hand. The CGS units most often encountered in acoustics are summarized in Table 1.2. Equivalence of units between the SI system (MKS) and the CGS system is shown in Table 1.3.

In this text, we will not attempt to use the English system or to

Table 1.1
SI Units.

Quantity	*Unit*	*Symbol*
Length	meter (metre)	m
Mass	kilogram	kg = 1,000 grams
Time	second	s
Electric Current	ampere	A
Force	newton	$N = kg \cdot m/s^2$
Work	joule	$J = N \cdot m$
Power	watt	$W = J/s$

Table 1.2
CGS Units.

Quantity	*Unit*	*Symbol*
Length	centimeter	cm = .01 meter
Mass	gram	gm
Time	second	s
Force	dyne	$dyne = gm \cdot cm/s^2$
Work	erg	$erg = 1\ dyne \cdot cm$
Power	watt	$W = 10^7$ ergs/s

Table 1.3
Unit Equivalents.

MKS		*CGS*
1 meter	=	100 centimeters
	=	1,000 millimeters
1 kilogram	=	1,000 grams
1 newton	=	10^5 dynes
1 joule	=	10^7 ergs
1 watt	=	10^7 ergs/sec

convert it to metric units. Certain branches of acoustics closely related to engineering still use the English system of measurement but this does not present significant difficulty in handling acoustic calculations. The fundamental unit of time, the *second*, is the same in all systems.

Linear Motion

Whenever scientists try to study a phenomena such as motion, they usually look at it in one of its simplest forms. In this way, the underlying principles of the phenomenon may be discovered and later applied to more complex forms of the same occurrence. The scientist may then study progressively more complex examples of the given happening until he has satisfactorily described it in those complex circumstances in which we encounter the phenomenon in our everyday lives. Such a progression from the study of very simple cases to the study of very complex cases may take centuries. Thus, we find the beginnings of the modern study of motion back in sixteenth century Italy in the workshop of Galileo Galilei. Galileo wished to study a simple example of the phenomenon of motion. He chose, as his simple case, the falling body. Lacking precise timing devices, he decided to slow the fall of the body to allow more careful study of how it fell. To do this, he allowed the body to roll down a smooth shelf that was elevated at one end. Such a tilted shelf is called an *inclined plane*.

Galileo's use of the inclined plane to study falling bodies is particularly important to us because it represents an approach to the study of physical phenomena that we will use repeatedly in this book. The direct study of falling bodies would have been impossible for Galileo because they fall too fast. He had no means of observing or measuring a phenomenon that happened as fast as a ball dropping from his hand to the floor. He therefore made the assumption that a ball falling freely to the floor and a ball rolling freely down an inclined plane show the same behavior. He assumed that the only difference in the two cases was the amount of time required for the ball to reach the end of its fall. By adjusting the inclined plane so that it was not very steep, the ball was made to travel slowly enough that the details of its motion could be observed and studied. What Galileo did was to develop a *model* of a freely falling body. In constructing a model, the scientist makes certain assumptions about the phenomenon he wishes to study. By then acting upon these assumptions, he develops a way of dealing with the phenomenon that is more amenable to scientific study. *Modeling* is often used to simplify the task of observing a phenomenon (as was the case with Galileo) or to allow the phenomenon to be more readily treated in mathematical terms. Often an occurrence in physics is too complex to be described by a simple mathematical equation. However, it is possible to make certain simplifying assumptions about the phenomenon that allow us to derive a mathematical equation that accurately describes most aspects of the occurrence for most conditions. We will encounter the use of modeling for purposes of mathematical description of phenomena later in the text.

Having been given a simple model for the study of falling bodies, we may now consider some useful aspects of this type of motion. At this point we are faced with a fundamental challenge of physics. How are we to go about the task of describing what happens when a ball rolls freely down an inclined plane? What can we say about this phenomenon more than the obvious fact that the ball starts at the top of the slope and stops at the bottom? A traditional approach to this challenge, and one which many feel is the essential function of modern physics, is to describe the phenomenon in numerical terms: to find means of ascribing a magnitude to various aspects of the occurrence. There are many numerical values that may be derived to characterize the ball rolling down the inclined plane. We will now consider these.

We can measure the length and height of the inclined plane. This will tell us the *distances* traveled by our freely rolling ball. Since we will be concerned in this book primarily with metric measurements, let us say that the length of the plank is 6 meters and its height at the raised end is 1 meter.

A second feature of Galileo's system that we can examine relates to *time*. We can observe how long it takes for the ball to roll to the bottom of the inclined plane. Thus, in the case we are currently examining, we may find that it takes 3 seconds for the ball to roll from the top of the inclined plane to its bottom.

One of the interesting features of physics is that basic quantities, as measured directly, may be combined mathematically. Such a combination allows us to develop other, more complex quantities that tell us more about what is happening than we knew from the original measurements. At this point in our examination of the inclined plane, we have measured the distance traveled by the ball (the length of the plank) and the time required to travel this distance. Time and distance are related by the concept of velocity. *Velocity* specifies the distance traveled by a uniformly moving body in a certain direction during a specific length of time. The physicist would say that velocity is a distance traveled in a certain direction per unit time. This statement can be written in the form of a mathematical expression: an *equation*. In mathematics, the symbol = represents the word *is* while the word *per* is represented by a division symbol. So, if we use the following symbols:

v = velocity,
d = distance, and
t = time,

we can mathematically define velocity by the equation

$$v = d/t\,. \tag{1.7}$$

If we now recall the values we measured for the inclined plane ($d = 6$ meters and $t = 3$ seconds), we can calculate the average velocity of the rolling ball by substituting these values in the velocity equation:

$$v = d/t = 6 \text{ meters/3 seconds}\,,$$

and

$$v = 2 \text{ meters/second} = 2 \text{ meters per second}\,.$$

Notice that we said 2 meters per second is the *average* velocity of the rolling ball. It is necessary to say this because of something we know about how objects move. Simple observation of moving bodies tells us that they do not begin moving suddenly at a high rate of *speed* (another word for velocity that does not imply direction but only the rate of movement.) Objects begin to move slowly and gradually increase their speed with time. When we divide the total distance traveled by the total time traveled, we get an *average velocity*. In talking with a friend about the distance to a nearby shopping center, you may say that the area is a half hour's drive from your home. In traveling the 20 miles to the shopping center you may drive along most of the time at 55 miles per hour and stop for three traffic lights. Your average speed, however, is distance (20 miles) divided by the time (1/2 hour) or 40 miles per hour, even though your car rarely traveled at that precise speed. Similarly, in the case of the inclined plane, the ball was standing still when we released it and started timing it. The physicist would say the ball was at rest or had zero velocity. As it began to roll down the slope it took some time to reach its average velocity of 2 meters per second. By the time it reached the bottom of the slope it was traveling faster than 2 meters per second, but its average velocity in rolling down the slope was 2 meters per second.

It is possible for us to examine this change of velocity demonstrated by the ball as it rolls down the inclined plane. To do this, we may look at the distance traveled by the ball in successive small periods of time. For our case, we will examine the distance traveled by the ball each second. The position of the ball at several time intervals after its release is shown in Figure 1.1. Examination of the figure shows that during the first second after release the ball travels from point A to point B—a distance of 0.6 meter. Thus, its average velocity during the first second is 0.6 meters per second. During the next second the ball travels from B to C—a distance of 2.0 meters. Its average velocity during this

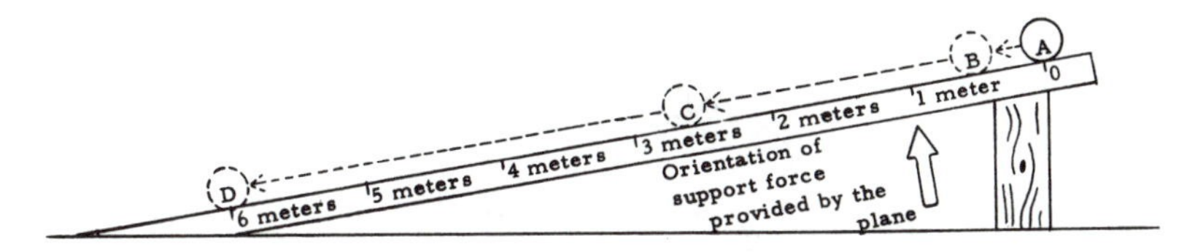

Figure 1.1 Movement of a ball down an inclined plane.
A - Starting position (rest position)
B - Ball position 1 sec. after release.
C - Ball position 2 sec. after release.
D - Ball position 3 sec. after release.

next second is, therefore, 2.0 meters per second. Similarly, its average velocity during the third second is 3.4 meters per second. The velocity of the ball is therefore changing throughout the time that the ball is rolling down the slope. A change in velocity of this sort is referrėd to by physicists as an *acceleration*. Note that the amount by which velocity increases is a constant of 1.4 meters per second for each second that passes. Thus, we say that the acceleration of the freely rolling ball is 1.4 meters per second per second or 1.4 meters/sec^2. Acceleration may be defined as the rate of change of velocity and it may be positive (increase in velocity with time—speeding up) or negative (decrease in velocity with time—slowing down). The reason why a freely rolling ball accelerates as it moves down an inclined plane will be discussed a little later in this chapter.

Another aspect of the inclined plane model that we can measure is the *weight* of the ball. While the measurement of weight seems like a simple task, it involves some rather sophisticated concepts in physics. We usually weigh an object by placing it upon a scale, so let us place our smooth ball on a scale and see what happens. When the ball is placed in the tray of the scale it causes the tray to move downwards by a certain amount—shown by arrow 1 in Figure 1.2. This downward movement of the tray pulls down on the chain supporting the tray. Pulling downward on the chain causes the needle on the dial of the scale to rotate clockwise as indicated by arrow 2 in the figure. The more we pull on the chain, the farther around the dial the needle moves. In this case, the needle reached the 1.6 on the dial and we report that the ball weighs about 1.6 kilograms. What we are really measuring in this case is the amount of downward pull exerted on the scale by the ball. The *weight* of an object, therefore, may be defined as the amount of downward push or pull it can produce. In physics, pushes and pulls have a specific

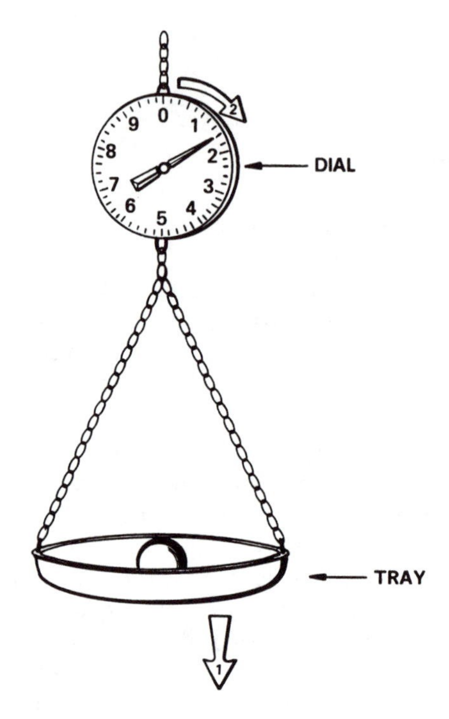

Figure 1.2. Weighing the ball on a scale.

name. They are called *forces*. The force we refer to as weight arises from a phenomenon called gravity. *Gravity* is the phenomenon whereby two things (the physicist would say "bodies") tend to fall towards each other (the physicist would say "attract"). Two factors control the magnitude of this attractive force: the distance between the two bodies (as distance increases, the attraction decreases and vice versa) and the *mass* of the bodies. Mass is a fundamental concept in physics. It may be defined as a measure of the amount of material present in a body. Thus, a lead bar of a given size has more mass than an aluminum bar of the same size. Similarly, a large lead bar has more mass than a small lead bar. Mass, therefore, depends upon the material of which a body is made and the volume of the body. The planet Earth, being made of massive materials and being large in volume (compared to other things around us) has considerable mass. An object near the Earth's surface, like Galileo's smooth ball, will be attracted by the Earth. Technically, then, if you drop the ball, both the ball and the Earth fall towards each other. Because the Earth is so much more massive than the ball, however, all we observe is the ball falling to Earth. Any Earth movement toward the ball would be far too small for us to see. In measuring the weights of various objects, we are measuring the force with which they fall to Earth. Since the Earth's mass remains the same and since the measurement location is essentially the same (the surface of the Earth) the measurement of the weight of an object on Earth depends entirely upon the mass of the object being measured. The moon is much less massive than the Earth. So if an object is moved from the Earth to the moon, its mass does not change, but it will weigh considerably less on the moon than it did on Earth. A man who weighs 156 pounds on Earth would weigh only 26 pounds on the moon, but his mass would be the same in either place. Thus, we may say that when we measure the *weight* of an object we are measuring the force with which it tends to fall towards some other object (usually the Earth). By contrast, the *mass* of the object defines the amount of material present in the object regardless of what other objects are near it.

In our consideration of weighing the ball, we have introduced the concept of force as a push or pull and we have noted that weight is an example of a force—the force due to the phenomenon of gravity. The concept of forces, especially as they arise from the interaction of bodies, was of particular interest to Sir Isaac Newton, the seventeenth century British physicist. He formulated some very important statements about forces and the motion of bodies which are known as *Newton's laws of Motion.* We will consider two of these important principles as we continue our study of linear motion. Newton's *first law of motion* states that an object tends to avoid any change in its existing state of motion. Thus, objects that are at rest like to stay at rest, while objects traveling along a straight line at a constant velocity just like to keep on going at that velocity (i.e., at a fixed speed in a given direction). Only the application of an outside force can make an object change its motional state. This tendency of an object to remain in its present motional state, in compliance with Newton's first law of motion, is known as the *inertia* of the body. When we release the ball at the top of the inclined plane, it would tend to stand still at the top of the slope due to Newton's first law, except for one thing—there is an outside force applied to the ball. It is the force due to gravity which acts on the ball and makes it begin rolling down the inclined plane.

Newton's experiments in mechanics led to his statement of the law of universal gravitation:

> Each particle of matter attracts every other particle of matter with a force which is directly proportional to the product of their

masses and inversely proportional to the square of the distance between them.

We can write this statement using symbols as:

$$F = G\frac{M_1M_2}{d^2} \tag{1.8}$$

where

F is the force of attraction,
G is the gravitational constant,
M_1 and M_2 are the masses of the particles, and
d is the distance between their centers.

In the special case where M_1 is the mass of an object such as a baseball and M_2 is the mass of the Earth, if d is the distance between the center of the Earth and the center of the object (the baseball), G becomes the familiar constant acceleration due to gravity of 9.8 meters per second per second (9.8 m/sec^2) and F is the force of attraction which we refer to as the *weight* of the object (in this case, the baseball).

Note that "weight" and "mass" are *not* the same. Using the metric system of measurement it is easy to understand the difference. The *mass* of an object is determined by the amount of matter in the object and is expressed in kilograms (kg). Using the acceleration due to gravity of 9.8 meters per second per second (9.8 m/sec^2), the *force* on the object which we refer to as "weight" is expressed in *newtons* with dimensions: kg m/sec^2. In more general terms, the principle describing the motion of bodies was first stated by Newton in 1683. This is the *second law of motion* which states that the acceleration, a, imparted to an object by an applied force (recall that a force has been defined as a push or pull) depends *directly* upon the force applied to the object and *inversely* upon the mass of the object. In other words, when the force is increased, the acceleration increases but when the mass is increased, the acceleration decreases. Stated mathematically:

$$a = F/M\ . \tag{1.9}$$

In considering the force we commonly call weight, we can observe a further aspect of the properties of forces. Weight is a force causing an object to tend to fall towards the center of the Earth. This is to say that the force due to gravity acts in a specific direction—downwards or towards the center of the Earth. This is a property common to all forces—they act in a specific direction. Whenever an object is pushed or pulled we can define a specific direction toward which the push or pull is oriented. In physics, any quantity which is associated with a particular direction is referred to as a *vector* quantity. By contrast, a quantity which has magnitude but is not oriented in a particular direction is called a *scalar* quantity. Mass, for example, is a scalar quantity. Keeping in mind that weight, being a force, is a vector quantity, let us take another look at the inclined plane.

The force due to gravity is pushing the ball towards the center of the Earth with an acceleration of 9.8 meters/sec^2, but the ball rolls down the slope at an acceleration of 1.4 meters/sec^2. This reduction in the acceleration of the ball compared to how hard the gravity force is pulling it arises from another principle of the motion of bodies. When we study the motion of bodies, we must keep in mind that the motion of the body will be in response to the total or net force acting on that body. A body will act in accordance with a given applied force, then, only to the extent that the force is unopposed. The inclined plane supporting the ball acts in partial opposition to the gravitational force on the ball. Since the plane is inclined, the support force it provides is not directly upwards but is perpendicular to the surface of the plane (see Figure 1.1). Thus, although the component of gravity force acting perpendicular to the plane is balanced, the component of the gravity force parallel to the plane is not balanced and therefore causes the ball to move along the plane at a much reduced acceleration and in a different direction from that associated with the gravity force.

When we roll the ball down the inclined plane, we may notice another effect worth considering. What happens when the ball reaches the bottom of the hill? Experience tells us that it does not stop but coasts on for a while even though it is on a perfectly horizontal surface. A horizontal surface is parallel to the Earth's surface and does not allow the ball to go down at all. Such a supporting surface would act in precise opposition to the force due to gravity both in magnitude and in direction. Under these conditions of *balanced opposing forces*, what we have said up to this point predicts that the ball will not move. We know from experience, however, that the ball will continue to roll once it reaches the bottom of the inclined plane. How does it do this? In order to answer this question, we must recall another characteristic of the moving ball. Recall that Newton's first law of motion is that an object tends to avoid any change in its existing state of motion (i.e., objects have inertia). Only the application of an outside force can make the object change its motion state. As the ball reaches the bottom of the hill it is traveling with some fixed velocity and it will continue to travel at that velocity until it is acted upon by some outside force. As we have defined the situation, another impossible occurrence has been predicted. Newton's first law dictates that the ball continue rolling at a fixed velocity forever. Experience tells us that this is not what will happen. The ball will coast for a while upon reaching the bottom of the slope but it will gradually slow down until it comes to rest. What force acts to stop the ball? The answer to this requires us to consider a final concept in motion physics—that of energy. In the study of motion we may define *energy* as an object's inherent capability to move. One form of energy is called *potential energy*. This is energy in a stored form. The ball at rest atop the inclined plane has potential energy due to its elevated position relative to the Earth's surface. It has the stored capability of moving to the bottom of the hill by virtue of its location and the phenomenon of gravity. Potential energy is often referred to as "energy of position". When the ball is released, the force due to gravity acts on the ball and sets it in motion. As the ball begins to move, it begins turning its potential energy into *kinetic energy*, the energy of motion. A body in motion has the inherent capability to remain in motion or to impart its motion to another object. This capability is referred to as kinetic energy. Once it reaches the bottom of the slope, the ball demonstrates inertia—a tendency to remain in motion at a fixed velocity as stated in Newton's first law of motion. The kinetic energy of the ball gives it the capability to remain in motion but it eventually stops. Why? As the ball rolls along the horizontal plane at the bottom of the slope, particles on the surface of the ball rub against particles of the plane. This rubbing is called *friction*. Friction or rubbing gives rise to heat. This is easily demonstrated by vigorously rubbing the palms of your hands together. The friction of one hand against the other generates heat. All matter may be characterized as being composed of millions of tiny, moving particles. When matter is heated, the motion of the particles of that matter is increased. It is this increased motion of the individual particles of an object that we call *heat*. Thus, in the case of friction, the kinetic energy of a moving object is converted into the kinetic energy of the particles of that object and of the surface against which it is rubbing. The phenomenon of friction, therefore, is one in which the kinetic energy of a moving body is used up in heating both the body itself and the surface upon which it is moving. So, friction is the force acting to cause the ball to eventually come to rest after reaching and passing the bottom of the inclined plane. Friction is also a significant concept for the study of acoustics and one that we will encounter again often.

CHAPTER 2
PERIODIC MOTION

So far we have developed some definitions for words that describe aspects of motion. By defining these words, we were able to study the fundamental phenomena associated with moving bodies. We will now expand our lexicon by defining some new terms and phenomena related to a special type of motion. The type of motion we will consider in this chapter has the property of repeating itself time after time. Since this motion is repeated periodically, it is called *periodic motion.* There are many examples of periodic motion in the world around us—the gentle rocking of a boat on the waves, the flapping of a shutter in a wind, the bouncing of a ball, the swinging of a person on a swing, the tapping of someone's foot or the clapping of his hands in time to some music, the movement of a pendulum in a grandfather clock. All of these motions are called periodic because they have one particular feature in common—they are all motions that are repeated at regular intervals. We will next consider the phenomenon of periodic motion in more detail.

THE PENDULUM

Recall what Galileo did when he wanted to carefully study some properties of motion. He selected a simple model of the motion he wanted to study—he examined a ball rolling down an inclined plane. We will do the same in our study of periodic motion. We will select an example of a simple motion that is repeated regularly in time as a model to help us in our study. One of the simplest and easiest to examine samples of periodic motion is the movement of a pendulum. The *pendulum* swings back and forth at precisely regular intervals. The pendulum's movement is so precisely periodic that it is used in clocks as a timing device. This is the model we will examine in our study of periodic motion.

Before considering the reasons for the phenomenon of the pendulum's motion, we will try to carefully describe what the pendulum's motion is like. Try to visualize, for a moment, the pendulum of a grandfather clock swinging slowly back and forth as it ticks off the seconds. What is happening?

We will describe a single pendulum swing beginning at the top of the swing (point A in the Figure 2.1). As the pendulum swings downwards towards point B, it moves faster and faster, gradually building up speed from a standing start at point A. It swings through the low point of the swing (at B) and begins to slow down gradually as it approaches point C where it stops altogether for an instant. Then, from a standing start at point C, it begins to move downwards again ever more rapidly through point B. After passing point B, the pendulum will gradually slow down until it stops for an instant at point A before beginning the cycle over again. Thus, we have said that a pendulum in motion repeats its motion very regularly; that it swings fastest through the midpoint of its movement (point B); and that it stops altogether upon reaching either endpoint of the swing (points A and C). Think for a moment about your own past experiences with a type of pendulum—the backyard tree swing. The exhilaration of fast motion comes at the time when the swing passes nearest to the ground—at the midpoint of

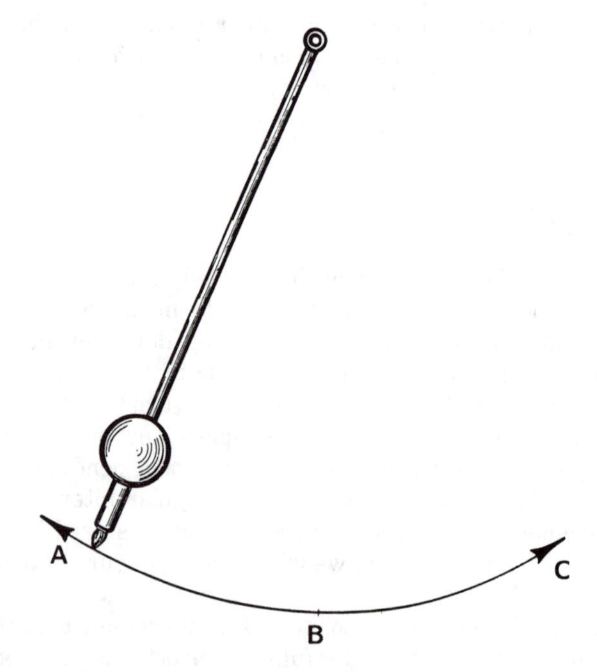

Figure 2.1. The pendulum in motion.

the swing. By contrast, the swing gradually slows down as it reaches higher and higher until, for a moment you hang suspended in air—motionless—at the endpoint of the swing. Then you start to swing downwards again ever faster as you approach the ground. Surely any swing or pendulum must go through this speeding up, slowing down, stopping, speeding up, etc., process. If the pendulum's motion were uniform, it would swing around in a complete circle rather than back and forth. We have, therefore, described the motion of a pendulum.

Let us consider for a moment what this description says about the nature of the motion shown by the pendulum. Recall that in Chapter 1 we described several ways of looking at motion; we can look at how far an object moves in a particular direction (*displacement*), we can look at the rate at which an object is changing its position or moving in a particular direction (*velocity:* displacement per unit time), and we can look at the rate at which an object's velocity is changing in a particular direction (*acceleration:* velocity per unit time). We will now look at the periodic motion of our pendulum in terms of each of these concepts—displacement, velocity and acceleration. In each case we will begin at the endpoint of a swing (point A in Figure 2.1). Unfortunately, even a pendulum moves too rapidly for us to study and measure its motion by simple observation. We would like to record its motion so we can study it more carefully. Suppose that the pendulum had an ink reservoir in it and a small opening at its bottom end through which the ink could slowly leak out. By filling the reservoir with ink, a tracing of pendulum motion could be drawn on a piece of paper held beneath the pendulum. If the paper were held still, a straight line would be drawn on the paper. The length of the line would correspond to the total extent in distance (endpoint to endpoint) traveled by the swinging pendulum. *Displacement* is the distance an object travels from its equilibrium position. Since the pendulum's equilibrium position is midswing (point B in Figure 2.1), the length of the line is twice the displacement magnitude or displacement amplitude.

The word *amplitude* is used here to mean magnitude. In acoustics, we often look at the amplitude of a motion. This may be a displacement amplitude, a velocity amplitude, or an acceleration amplitude and it is necessary to specify what type of amplitude you mean when you are talking about the "amplitude" of a motion. Thus, the length of the line is the absolute total of the displacement amplitudes in both directions from midswing position. Since displacement is the distance traveled from equilibrium, the displacement amplitude is the distance from the midpoint of the line to either end—that is, half the length of the line. In defining the movement of the pendulum, we will arbitrarily say that movement to the right of the midswing position is in the negative displacement direction while movement to the left of midswing is in the positive displacement direction. In this way, we can know the direction of the displacement from midswing by knowing whether the displacement amplitude has a positive or a negative sign at that instant in time.

Now, if paper were moving past the point of the pendulum at a constant rate, a wavy line or a wavy string of dots would be drawn by the pendulum's swing (that is, at right angles to the displacement amplitude line we just drew) as shown in Figure 2.2. The wavy line drawn by most objects that move periodically has a characteristic shape similar to that shown in Figure 2.3. This line is really a graph of the displacement amplitude of the pendulum as time passes. If the paper is

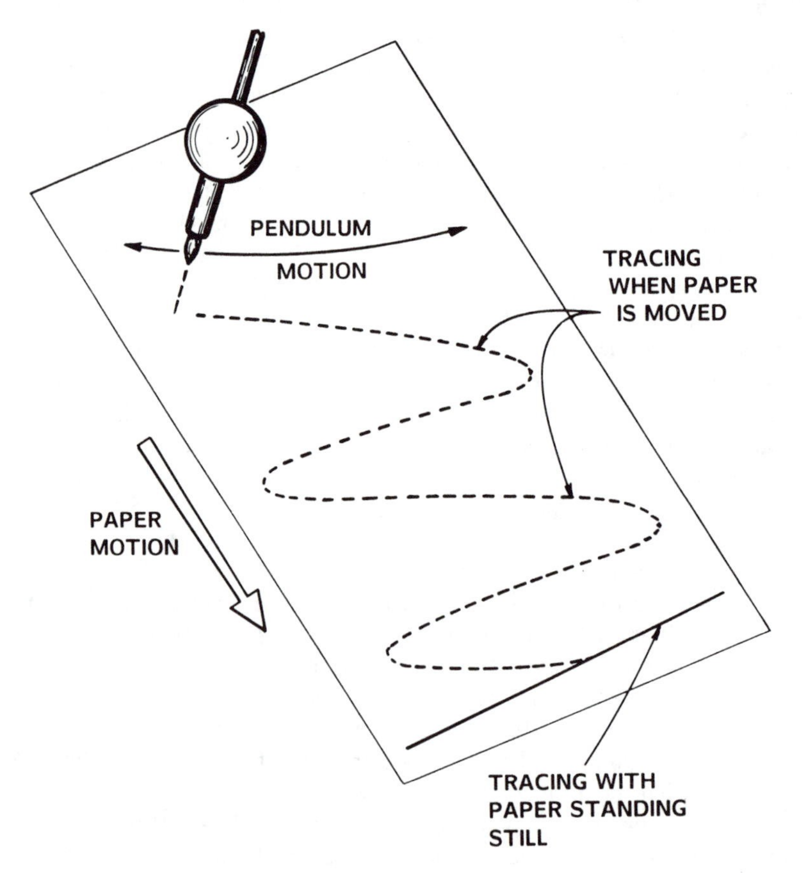

Figure 2.2. The displacement of a pendulum as a function of time.

moved beneath the pendulum at a constant speed, then every point on the wavy line tells us where the pendulum was at a particular instant in time. This type of graph is used often in acoustics. It is often called an amplitude versus time plot. The horizontal axis of the graph represents elapsed time; the time since we began looking at the motion. The time starts at some arbitrary starting point and we measure how much time has gone by since the start. It could be measured in seconds. The vertical axis of the graph tells how far from the midpoint of the swing the pendulum has swung. When a periodic motion gives rise to a graph such as that shown on Figure 2.3 (having smooth transitions throughout), we say that the motion is a *simple harmonic motion*. In physics, this type of graph is often called a *sine curve* since the instantaneous displacement amplitude (on the vertical axis) is a *sine function* of the time values (on the horizontal axis). The word *function* is used in physics to indicate a dependent relationship between two or more variables. A *variable* in this context is a quantity that may assume any one of a set of values. For example, the amount of perspiration a person produces is a function of the temperature and humidity in his environment. The amount of perspiration produced depends upon the temperature and upon the humidity of the environment. In the case we are considering, the instantaneous displacement amplitude changes in its magnitude as time goes on, in the same way that the sine of an angle varies with the size of the angle. Thus, when the instantaneous amplitude of a periodic

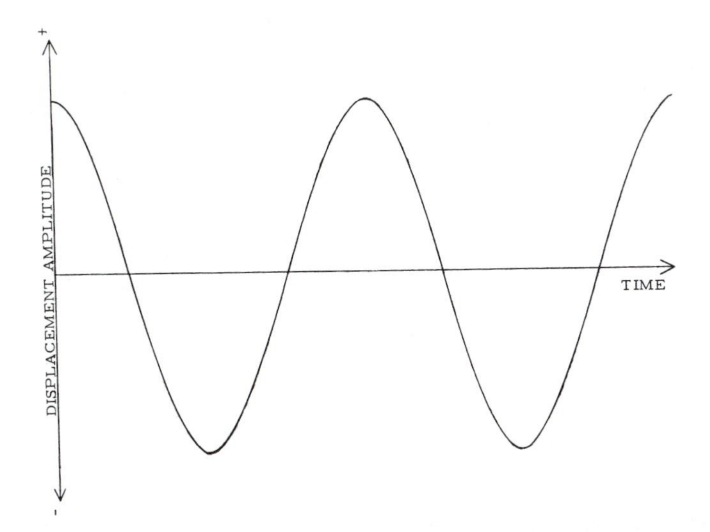

Figure 2.3. Amplitude versus time plot of simple harmonic motion.

motion is a sine function of time, the motion is called a simple harmonic motion.

It will now be useful to define some terms that are used in describing the particular characteristics of a given simple harmonic motion. The word *cycle* is used to denote a single repetition of a periodic motion. Thus, for the simple harmonic motion we are looking at, one cycle would be the motion from rest through a positive amplitude peak through rest to a negative amplitude peak and back to rest. This is shown in Figure 2.4. Note that the first peak may be positive or negative in defining one cycle. In fact, we may even begin from some point other than the rest position so long as the motion represents one and only one repetition of the periodic activity being examined.

In describing the simple harmonic motion of a particular system, we might want to describe the relationship between that motion and the passage of time. One way to define this is to say how long it takes for one cycle of a simple harmonic motion to happen. The time required for one cycle of periodic motion is called the *period* of the motion. The period of a motion may be read directly from a time-amplitude plot of

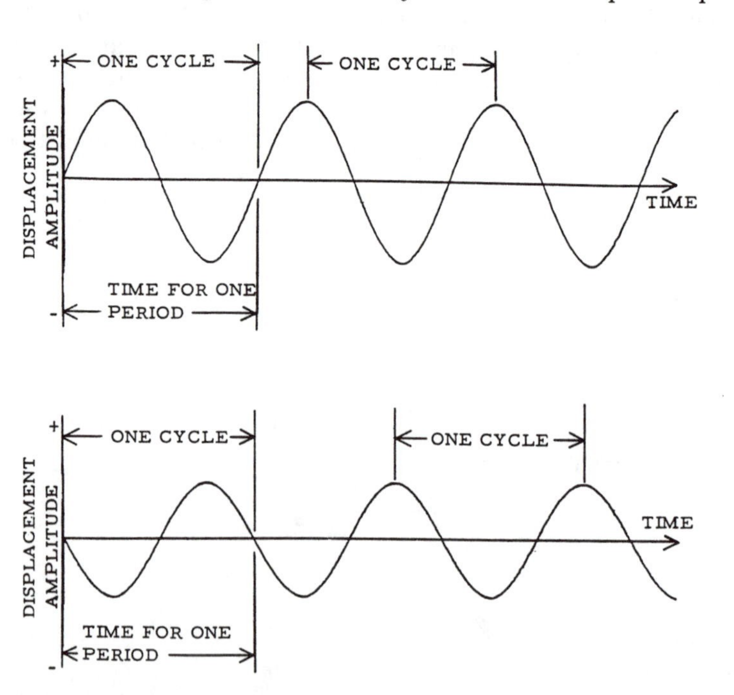

Figure 2.4. Examples of simple harmonic motion amplitude versus time plots of displacement.

a motion. By noting (as in Figure 2.4) the region occupied by one cycle on the graph, we may find that one cycle takes up a specific distance along the time axis. This distance represents the time required for one cycle or the period of the motion. This concept is used in many areas—the definition of a 24 hour day is the time required to go through one complete cycle of daylight and darkness. Thus, 24 hours is the period of the daylight/darkness cycle here on earth. If we define sunrise and sunset as equilibrium points in the cycle, we may then arbitrarily assign a positive value to the condition of light and a negative value to the condition of darkness. Note that this positive/negative value is assigned arbitrarily. It is only done so that we may know the direction a phenomenon is heading with reference to the equilibrium state. This arbitrary assignment of positive and negative values to specific directions a phenomenon may take relative to an equilibrium condition is commonly done in physics and is something the student will encounter again and again.

Another way of defining the relationship of simple harmonic motion and time is to specify the number of cycles that occur in a specific time duration. In our daylight/darkness example, if we use the week as a duration of time we may see that the daylight/darkness cycle is repeated seven times in that length of time. We might say that daylight/darkness has a frequency of occurring seven times in one week or more simply that it has a frequency of seven cycles per week. *Frequency* is the number of repetitions of a periodic phenomenon that occur in a given duration of time. In acoustics, the phenomenon we usually examine is periodic motion, and the basic duration of time is one second. We measure the period (T) of a motion in seconds. The frequency (f) is then given, simply, as the inverse of the period. (Recall that the inverse of a value is 1 divided by that value.) This relationship may be written mathematically

$$f = \frac{1\ (\text{cycle})}{T\ (\text{sec})}\ . \tag{2.1}$$

Thus, if a motion has a period of one second, it has a frequency of one cycle per second. The word *hertz* is used to denote the phrase "cycle(s) per second". Thus, if a motion has a period of one second (sec), it has a frequency of one hertz (Hz). Furthermore, if a motion has a period of $1/100^{\text{th}}$ of a second, it has a frequency of 100 Hz.

In the speech and hearing sciences, we are concerned with those sounds that can be heard. These sounds have a range of periods from $1/20^{\text{th}}$ sec to $1/20{,}000^{\text{th}}$ sec. That is, we are interested in those sounds occurring in the frequency domain from 20 Hz to 20,000 Hz.

In studying our pendulum so far, we have examined only its displacement amplitude as a function of time. Let us now consider its velocity amplitude as a function of time, beginning again at the endpoint of a swing (point A of Figure 2.1). Recall we said earlier that the pendulum is actually stopped at the endpoint of the swing for an instant of time.

It then begins to move faster and faster as it swings downward (we will use from left to right as negative direction for velocity since left is the positive displacement direction) until it reaches a velocity maximum (in the negative displacement direction) as it swings through the equilibrium position. Continuing to swing, now upwards, it gradually slows down until it stops at the endpoint of the swing (point C in Figure 2.1) for an instant of time. It then returns (in the positive, right-to-left direction) increasing velocity to a maximum velocity at the equilibrium position and then slowing down in its gradual upward movement until it reaches a zero velocity at the swing's endpoint (point A in Figure 2.1). A graph of this velocity amplitude is shown in part (b) of Figure 2.5. For comparison, the displacement amplitude versus time relationship for the same pendulum motion is given in part (a) of Figure 2.5. Note that while the two functions are very similar (they are both sinusoidal functions of time), they go through their maximum and minimum points at different instants in time. When displacement is at a positive peak (the pendulum is instantaneously stopped at point A of Figure 2.1), the velocity is zero since the pendulum is stopped for a brief instant in time. By contrast, when the displacement is zero (the pendulum is swinging through rest position at point B of Figure 2.1), the velocity is at its greatest. Also note that the length of one cycle on the time axis (that is, the period) of the displacement amplitude is the same as that for the velocity amplitude function. Thus, we may say that while the two phenomena do not coincide precisely in time, they have the same period and they are, therefore, at the same frequency. When two phenomena are at the same frequency but do not precisely coincide in the time domain, we say that they are *out of phase* with one another. Conversely, when two phenomena are at the same frequency and do coincide precisely in timing (going through positive maxima, zeroes and negative maxima at the exact same instants in time) we say that the two functions are *in phase* with one another.

When two phenomena are out of phase with one another, it is often useful to be able to specify the precise amount by which they differ in *phasing*. Look carefully now at parts (a) and (b) of Figure 2.5. You may notice that, at any instant in time, the part of the periodic cycle being shown by the velocity graph is that part which occurs 1/4 cycle before

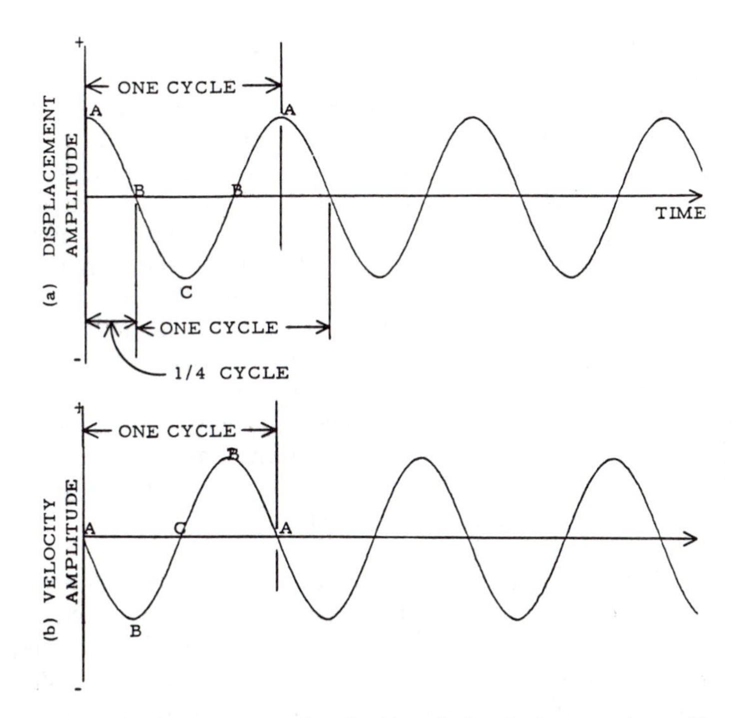

Figure 2.5. Displacement and velocity of simple harmonic motion as a function of time.

the motion component shown by the displacement graph. If we look at the displacement graph starting 1/4 cycle after the beginning of the motion, we see that it is precisely the same as the initial cycle of the velocity graph. We may, therefore, say that the displacement property of a simple harmonic motion *lags* 1/4 cycle behind its velocity component. For mathematical purposes, the time required for one cycle is often divided into 360 degrees (360°). Thus, we often encounter the statement that in simple harmonic motion velocity leads displacement by 90° (i.e., 1/4 of 360°). All this says is that the maximum (or minimum) of the velocity amplitude of such a motion will occur 1/4 of a *period* before the corresponding maximum (or minimum) of the displacement amplitude of that motion.

Now let us look at the acceleration component of our pendulum's motion. Recalling that acceleration is the rate of change of the velocity, we can determine it by knowing how the velocity is changing during any small interval of time. At the endpoint of the swing (point A in Figure 2.1) the velocity begins to increase as it moves downward. At first the velocity change is large as we go from standing still to the downward swing (so acceleration is large). Then the pendulum gradually approaches its maximum velocity (seen at point B in Figure 2.1). The increase in velocity is more and more gradual as maximum velocity is approached, so acceleration decreases as point B of Figure 2.1 is approached. Then, as the pendulum swings through its midpoint, velocity is a constant maximum, so acceleration is zero at point B in Figure 2.1. Continuing through point B of Figure 2.1, the pendulum begins to change velocity, at first gradually, then ever more rapidly as the pendulum approaches point C of Figure 2.1, so acceleration increases while approaching the endpoint of the swing. When the pendulum reaches point C (Figure 2.1), the pendulum's velocity goes to zero for an instant while the pendulum prepares to begin moving back downwards. Recall that velocity is a vector quantity having both magnitude and direction. While the velocity magnitude is zero for an instant of time at point C (Figure 2.1), the pendulum is, at that instant, changing its direction of motion and so velocity is really changing. This *change* in velocity is a very large one since the motion turns completely around at this point and begins going in the opposite direction. In fact, at point C (Figure 2.1) the acceleration reaches a maximum. As the pendulum then swings through Point B and returns to point A (Figure 2.1), the same acceleration phenomena as previously described are now encountered going in the opposite direction. The acceleration component of the pendulum's motion is shown in an amplitude versus time plot in part C of Figure 2.6. For comparison, the displacement and velocity amplitude versus time graphs of the simple harmonic motion are shown in parts A and B, respectively, of that figure. Examination of these graphs show that the acceleration leads the velocity in time by 1/4 cycle or 90°. Comparing acceleration to displacement shows that these two phenomena go through the zero axis at the same instant in time but they are moving in opposite directions from the zero axis (i.e., displacement has a *negative maximum* at the same instant in time that acceleration has a *positive maximum*). This condition indicates that acceleration and displacement of a simple harmonic motion are half a cycle or 180° (half of 360°) out of phase with one another. As we may see from examining Figure 2.6, periodic motion may be examined by studying either displacement, velocity, or acceleration aspects of the motion. We need only keep in mind that these aspects of motion are not in phase with one another. Some of the implications of these phasing differences will be discussed in a later chapter.

Up to now we have looked at the motion of a pendulum without really considering the forces acting to give rise to its simple harmonic

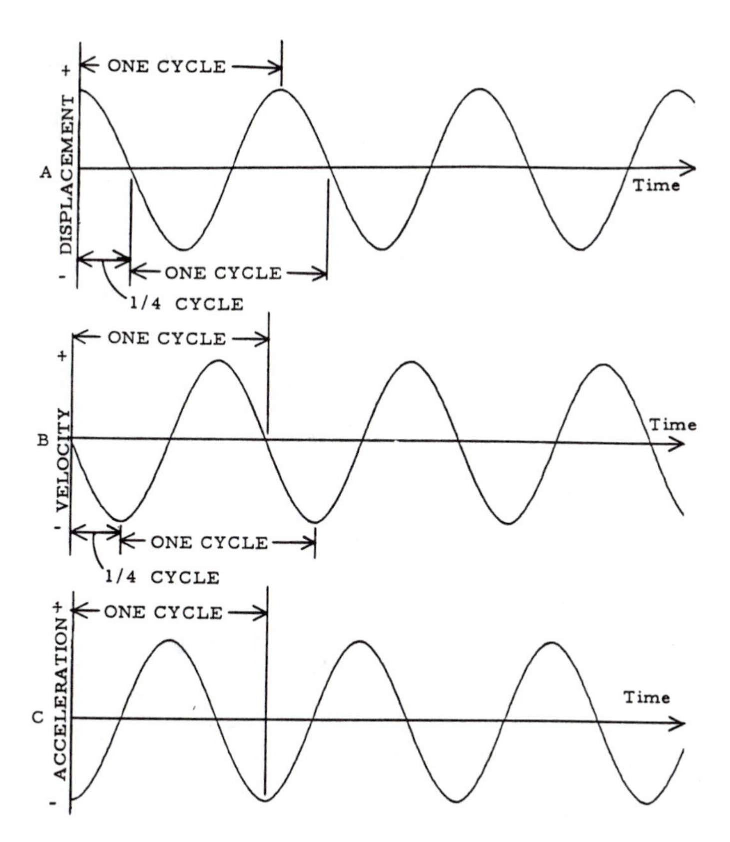

Figure 2.6. Time relationships of displacement, velocity and acceleration for simple harmonic motion.

motion. Let us think about the pendulum in terms of force concepts. The movement of the pendulum is limited by the pendulum arm. Since the arm is fixed in length, the pendulum's movement describes an arc (a part of the circumference of a circle) so that as the pendulum falls its movement becomes less and less downwards and more and more horizontal. In fact, this phenomenon may be explained in terms of the forces involved. The pendulum arm exerts a force (a pull) on the pendulum mass and this force is directed towards the attachment point (the pivot point) of the pendulum. The magnitude of this force is equal to the magnitude of the weight (gravitational force) exerted upon the mass in the pendulum (point D in Figure 2.7). We can see this if we consider the pendulum when it is not swinging but is at rest with the mass hanging straight down from the pivot point (see Figure 2.7). If

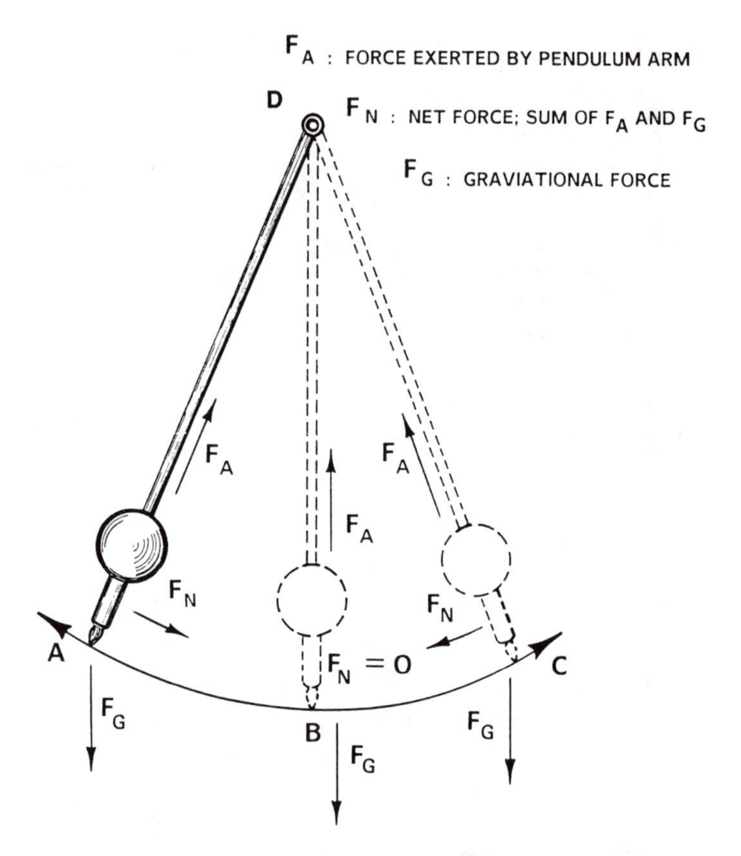

Figure 2.7. Forces in pendulum motion.

the magnitude of the force arising from the pendulum arm were greater than that of the gravitational force, the pendulum mass would be pulled upward (i.e., shortening the arm) and would not remain at rest. Conversely, if the magnitude of the pendulum arm force were less than that of gravity, the mass would be pulled downward (i.e., lengthening the arm). The fact that the mass remains motionless tells us that the two forces are balanced and in opposition to one another. This is to say that the net force on the pendulum's mass is zero, as noted in Figure 2.7. This is not the case in the instant of time when the pendulum is stopped at one endpoint of its swing (e.g., pint A in Figure 2.7). In another instant of time, the pendulum will begin the swing diagonally downwards (towards point B). If it is about to begin moving in a particular direction, Newton's first law tells us that the net force acting on the mass must be greater than zero. In fact, you can see from the diagram of forces in Figure 2.7 that the net force on the pendulum mass is not zero. It is true that the gravitational force is equal in magnitude to the pendulum arm force (as was the case at point B) but now they are no longer oriented in opposite directions from one another. Thus, they no longer add to zero net force on the pendulum mass. In fact, it may be seen that the farther the pendulum is from its midswing position, the more the two forces deviate from being in direct opposition from one another. The more the two forces deviate from being in direct opposition to one another, the greater the net force on the mass. Therefore, the further the mass is from its midswing position, the greater the magnitude of the net force acting on it. The direction of the net force acting on the pendulum mass will be halfway between the directions of the two forces acting on it (so long as these two forces are equal in magnitude—as is the case in a pendulum). This means that the net force acting upon the pendulum mass will always be oriented towards the resting or midswing position of the pendulum. Because this net force acting on the pendulum is directed in such a way as to restore the pendulum to its rest position, we refer to this net force as the *restoring force* for the pendulum. As the pendulum moves from position A towards position B, the magnitude of the restoring force is continually decreasing because the two forces acting on it (F^A, the pendulum arm force, and F^G, the gravitational force) are moving ever closer to being in direct opposition to one another. The velocity of the mass is continually increasing as it moves from A to B but the rate of velocity increase (i.e., the acceleration) is lower as the mass approaches the equilibrium position. This decrease in acceleration as the net restoring force decreases is in agreement with Newton's second law of motion.

We may also consider this phenomenon in terms of the energies present. When the pendulum's mass is displaced from its equilibrium or rest position, the net force acting on it gives potential energy to it. As the pendulum moves closer to the equilibrium position, its potential energy is converted into energy of motion—kinetic energy. Thus, the potential energy is decreasing while the kinetic energy is increasing. At midswing (point B of Figure 2.7) the motion is completely horizontal.

The next question we face is: Why does it keep going? At midswing, the restoring force has moved the pendulum to a position that is most downward—nearest to the center of the earth—and since the forces acting on the pendulum balance perfectly to a zero net force at this position, the pendulum would like to stop in that position. Experience tells us, however, that the pendulum continues to move past the midswing position. The phenomenon that keeps the pendulum moving past midswing is the same one that caused the ball in our previous discussion to roll past the bottom of the inclined plane—Newton's first law of motion. Once a pendulum of a specific mass is set into motion, that mass wants to continue moving. This tendency, demonstrated by a moving body, is directly proportional to the mass of the body. Thus, the mass of the pendulum gives rise to a tendency for the pendulum to pass through its midswing position and on towards the endpoint of the swing (point C in Figure 2.7).

As the pendulum continues to swing, the arm, since it is fixed in length, causes the pendulum to move upwards as it moves away from the equilibrium position. The force due to gravity now begins to be oriented in a direction that is not in direct opposition to the force arising from the pull of the pendulum arm. The resulting net force (a restoring force) is now oriented in the opposite direction from the motion of the pendulum and it acts to slow the pendulum down as it swings upward. In effect, the restoring force is acting in opposition to the tendency of the mass to remain in motion. As the pendulum moves upward towards point C (of Figure 2.1), the energy in the pendulum changes form again. As the pendulum swung from point A to point B, its potential energy (energy of position) was transformed or converted into kinetic energy (energy of motion). By expending its kinetic energy, the pendulum is next able to move from point B to point C overcoming the restoring force until, once at point C, the kinetic energy has been completely converted back into potential energy and the pendulum begins its return swing to point B. As the pendulum swings, therefore, its energy is constantly changing form, from potential to kinetic and back again. It is important to keep in mind that the total energy in the system (the sum of kinetic and potential energies) is constant and that only the relative magnitudes of each form of energy is changing. The potential energy arises from the restoring force by virtue of the *position* of the pendulum mass with respect to its equilibrium position. Thus, as the moving body gets farther away from the equilibrium position, the energy arising from the restoring force is increased until at the endpoint of the swing all of the energy is in the potential form.

As the body (the mass of the pendulum) begins to move towards the equilibrium position B, the potential energy is converted into kinetic form. Recall that kinetic energy is the energy of motion and that the pendulum's time rate of motion (velocity) continues to increase as it approaches the midswing position. At midswing the system's energy is completely converted into the kinetic form, since this is the point of maximum velocity. The pendulum then uses this kinetic energy to move through the midswing position and counteract the restoring force. The tendency of a mass to continue in motion once set in motion is the phenomenon described in Newton's first law of motion; this tendency is in proportion to the mass of the body. Thus, it may be seen that the mass present in the system is proportional to its kinetic energy.

The moving pendulum, therefore, continues in motion by transforming its energy back and forth between the kinetic and the potential form. This transformation of energy is sustained by the operation of two phenomena on the system—the restoring force and the inertia acting in accordance with Newton's first and second laws of motion. The restoring force gives rise to potential energy and tends to make the mass move towards the equilibrium position. The inertia is the tendency of the mass to obey Newton's first law of motion and leads to the fact that the pendulum possesses kinetic energy thereby tending to make the mass of the system move past the equilibrium position and towards the endpoints of the swing.

These relationships may be found in all systems that demonstrate simple harmonic motion. In a pendulum, the inertia is proportional to the mass of the pendulum itself. As we will see in studying many acoustical systems, the inertia is defined as a phenomenon arising from the mass of the moving body in a system. Restoring forces, however, are variable in origin and will not necessarily arise from a single factor. In fact, in most acoustical systems the restoring force is not related to the force due to gravity, but arises from the elasticity of matter, a concept which will be considered in a later chapter.

We have said that periodic motion will not exist without an interaction of the two phenomena—inertia and restoring force. Think for a moment about a pendulum with no inertia. If such a pendulum could be devised, it could not swing. The first time it reached the equilibrium position, the restoring force would act to keep it still and, without inertia to counteract the restoring force, the pendulum would stop. It would not demonstrate periodic motion. Now, in contrast, think of a pendulum with no restoring force. This could be done by moving the pendulum off the Earth into free space where there would be no gravity force. In this case, once the pendulum was displaced, the inertia would cause it to stay in its displaced position. There would be no force acting on it to cause it to move. With no restoring force to counteract the inertia, the pendulum would become motionless as in the earlier case and we would not see the periodic motion we had back on Earth.

Thus, in order for a system such as a pendulum to demonstrate periodic motion, both an inertia tendency and a restoring force are necessary. Furthermore, these phenomena must interact so that the resulting potential and kinetic energies can trade off control of the system.

Our experience with pendulum motion tells us that a pendulum does not swing forever. It tends to decrease the amplitude of its swing gradually so that, in time, it comes to rest at the equilibrium position. Our description of the forces acting on a pendulum is thus incomplete. The description to this point (taking into account inertia tendency and restoring force) is sufficient to explain why a pendulum stays in harmonic motion but it does not explain why the pendulum eventually stops this motion. Considering Newton's first law of motion, we must conclude that we need another force; one that will explain why the pendulum stops. Recall our discussion of the ball rolling beyond the bottom of an inclined plane. The force that caused that ball to eventually come to rest is the same one that causes the pendulum to come to rest. This is the *frictional force* encountered by a moving body. The pendulum, in swinging, pushes air out of its way. This air slides over the body of the pendulum and in rubbing against it causes heat to be generated. Recall from our previous discussion of frictional forces that any process of rubbing giving rise to heat generation uses up energy. This frictional force uses up the energy of the pendulum and it is therefore referred to as *dissipative energy*. This activity reduces the amplitude of the pendulum's swing and in time will dissipate all of the pendulum's energy, causing it to stop. In any real pendulum, a second source of friction is the rubbing that takes place at the hinge or pivot attachment at the top of the arm (point D in Figure 2.7). When the pendulum is suspended, its arm will encounter frictional forces at the hinging point and these will further act to dissipate energy and reduce the swinging of the pendulum. Figure 2.8 shows a plot of displacement amplitude as a function of time for a pendulum showing the effects of frictional force. Note that the friction results in a progressive decrease in the pendulum's displacement amplitude as time goes on. Also observe that the frequency of the periodic motion shown by the pendulum is about the same with friction as it was without friction. (Note: In reality, the addition of a frictional force slows down the movement of the pendulum *slightly*, thus, increasing its period. Increasing the period of the pendulum's motion will decrease the frequency of its motion, since $f = 1/T$. In the real world, this change in the frequency of the pendulum's periodic motion due to friction is quite small for most cases. The significant effect of adding the frictional force to our pendulum is the effect on the amplitude of motion components.) Thus, the primary effect of the frictional force is to progressively diminish the amplitude of the pendulum's periodic

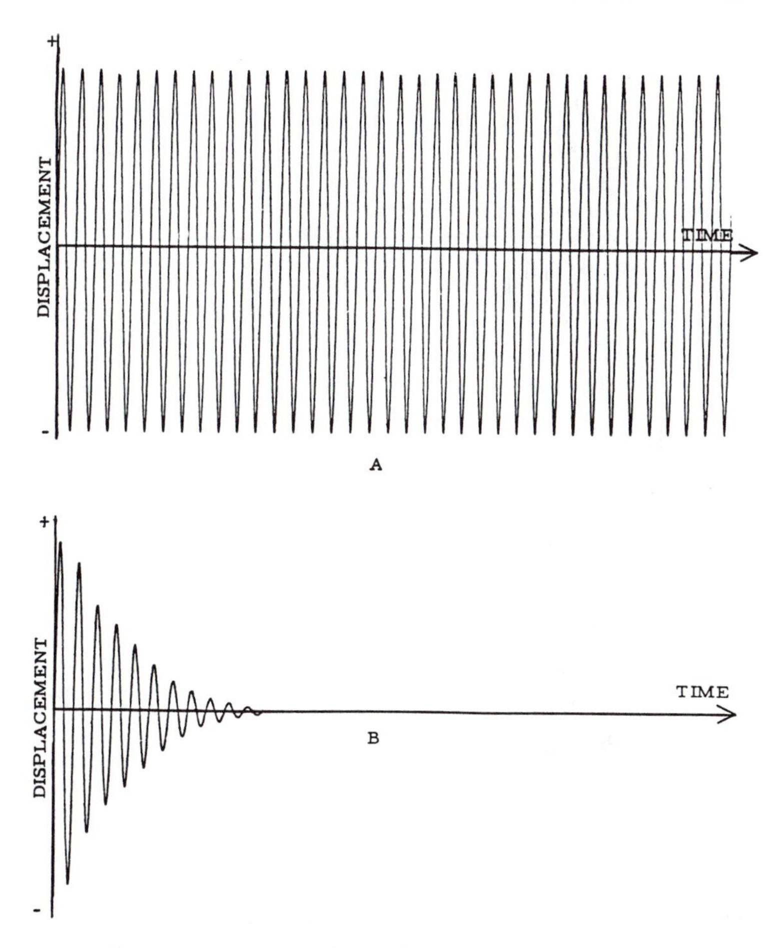

Figure 2.8. Motion of a pendulum (A) without friction and (B) with frictional force.

motion. This harmonic motion with amplitude decrease is referred to as *damped motion.* In acoustics we often characterize the magnitude of the friction acting on a system by defining the amount of *damping* displayed by the motion of the system. Figure 2.9 shows the displacement amplitude of a pendulum as a function of time showing varying amounts of system damping. Note that as the damping is increased, the progressive diminishing of the displacement amplitude occurs more rapidly. One effect of this is that the total time during which a pendulum will swing decreases as damping increases. If the damping is further

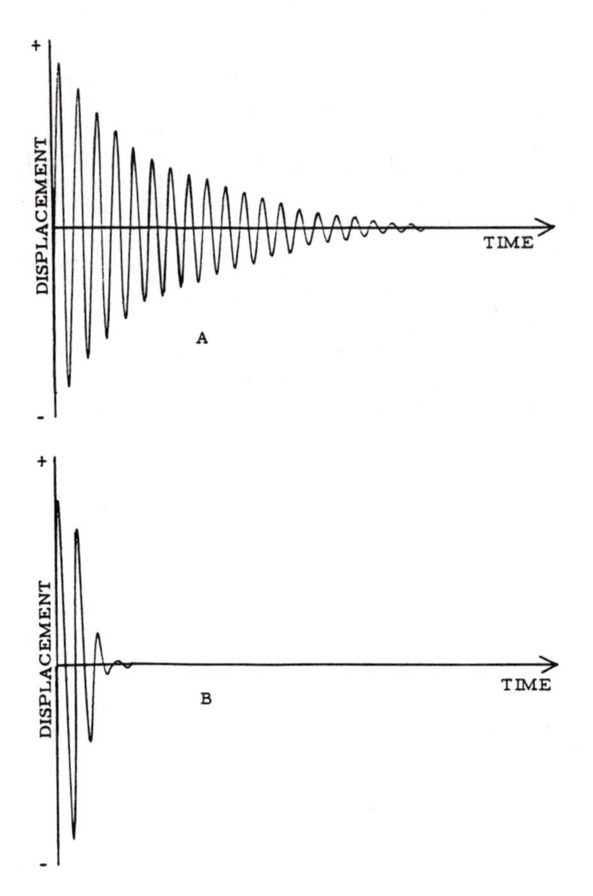

Figure 2.9. Motion of a pendulum showing (A) small amount of damping and (B) large amount of damping.

increased, we will eventually reach a condition where all the energy will be dissipated by the time the pendulum reaches the rest position. Under this condition, the pendulum will stop moving at the rest position. When a system demonstrates this phenomenon we say the system is showing *critical damping*. The displacement amplitude of a critically damped pendulum as a function of time is shown in Figure 2.10. The amount of damping in a system can be expressed by measuring the time required for the oscillations of the system to decay a standard amount. This decay time is called the *decay modulus* and is defined as the time required for the oscillations to decay to $1/e$ = .37 of the steady state or initial amplitude. Damping is a necessary aspect of many acoustical

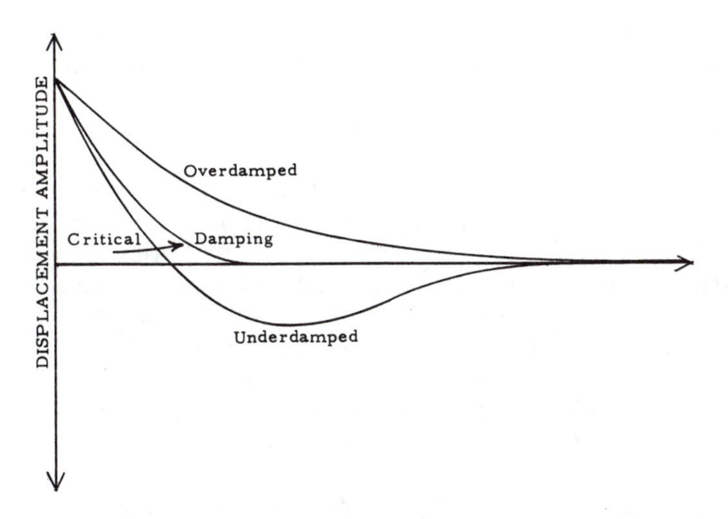

Figure 2.10. Motion of a pendulum showing critical damping and over-damping.

systems where it is desirable for periodic motion to be stopped soon after it begins so the system will be ready to deal with the next signal. The movement of the basilar membrane of the inner ear demonstrates nearly critical damping. If it were not so damped then when a sound entered the hearing mechanism, we might hear it echoing or ringing for many minutes after it occurred. Such a situation would interfere with listening to later sounds and would probably cause a good deal of annoyance.

The decay modulus is related to the system properties by the formula:

$$\tau = \frac{1}{\beta} = \frac{2M}{R}$$

where: τ = decay modulus
M = the moving mass
R = the mechanical resistance (friction)

System is critically damped when $\beta = 2\pi f_0$

The Spring-Mass System

As noted in the introduction to the previous section, the pendulum is not the only example of periodic motion that we may study. In acoustics, one of the most useful models of periodic motion is the spring-mass system. It was not introduced initially in this chapter because it is not a system with which we all have obvious experience as is the pendulum. The spring-mass system, like the pendulum, is a simple model of a specific type of periodic motion. This system is composed of two springs with one end attached to a fixed support. The other end of each spring is attached to a weight or mass. The system is shown in Figure 2.11. We will briefly consider the motion of this system.

This description will examine one cycle of the spring-mass system's periodic motion beginning at the one endpoint of the motion (point A in Figure 2.11). As the mass moves progressively towards point B, it moves faster and faster, gradually building up velocity from its standing start at point A. It moves through the equilibrium position (at point B) and begins to slow down gradually as it approaches the other endpoint of its movement at point C. Upon reaching C, the mass stops altogether for a brief instant. Then, from a standing start at point C, it begins to move back again ever more rapidly through point B. After passing point B, the mass will gradually slow down in its motion until it stops for an instant at point A before beginning the cycle over again. Compare this description of motion in the spring-mass system with that the pendulum motion in Figure 2.1. Clearly, the two systems display very similar types of motion. As with the pendulum, the motion of the mass in a spring-mass system is repeated very regularly; is fastest at the midpoint of the movement (point B) and is stopped altogether for an instant at each endpoint of the movement (points A and C).

Since the motions of these two systems are very similar, it could be concluded that many of the fundamental features of pendulum motion would also be present in the motion of the spring-mass system (*S-M*). This is, in fact, the case. The time function of the displacement amplitude for the *S-M*'s mass is sinusoidal like that of the pendulum. For an example, see Figure 2.3. As with the pendulum, the velocity amplitude

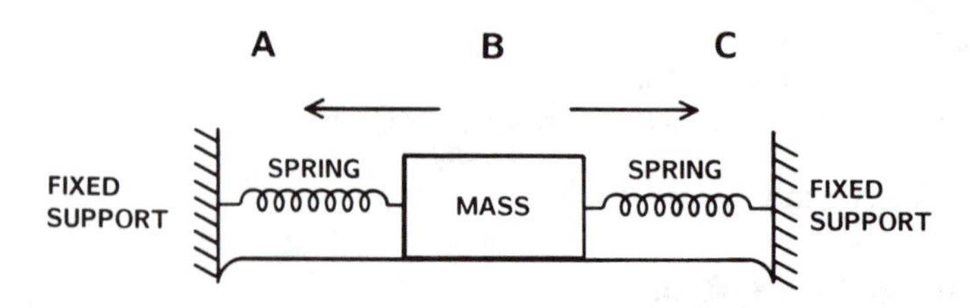

Figure 2.11. The spring-mass system shown in a schematic diagram. Points A, B and C denote specific position of the system's mass at different times during its movement.

of the *S-M* mass is a sinusoidal function of time and it has its greatest amplitude at the equilibrium position (point B) where the displacement amplitude is zero. Similarly, the velocity of the *S-M* mass is zero at the endpoints of the movement (points A and C) where displacement amplitude is a maximum. Thus, for a spring-mass system in free periodic motion, the displacement amplitude lags 90° behind the velocity amplitude as shown in Figure 2.5. Finally, the acceleration of the *S-M* mass is greatest when the direction of its motion is changing (at points A and C) while the acceleration is a minimum when the mass achieves its maximum (but constant) velocity in a given direction at the equilibrium position (point B). Thus, acceleration amplitude (a sinusoidal function of time in the *S-M* system) leads velocity amplitude by 90° and is 180° out of phase with the displacement amplitude, as shown in Figure 2.6. As you can see, the fundamental relationships among the three components of motion are the same for a spring-mass system in free periodic motion as they are for a pendulum in free periodic motion.

The phenomena which act on the mass of a spring-mass system to keep it in simple harmonic motion are inertia and restoring force. As with the pendulum, the restoring force gives rise to the energy of position—potential energy. The restoring force in a spring-mass system arises from the spring. The spring would like the *S-M* mass to be situated at a specific location (the equilibrium position at B). When the mass is away from that position (i.e., towards A), one spring is compressed more than it wants to be, while the other is stretched more than it prefers and they exert a restoring force tending to push the mass back towards Point B. As the mass moves closer to point A, the restoring force of the spring converts more and more of the system's energy into potential form. Once point A is reached, the kinetic energy of the system is completely converted into potential form and the *S-M* mass stops for an instant before beginning its return trip to the equilibrium position. As the *S-M* mass returns towards B, its potential energy is slowly converted back into a kinetic form. Just as we have seen in the case of the pendulum, the mass of the *S-M* system gives rise to its inertia, the factor controlling the system's kinetic energy—energy of motion. Due to inertia, the system operates to convert the system's energy to kinetic form, resulting in the *S-M* mass moving through the equilibrium position at B and on towards C. As the *S-M* mass moves towards C, the spring is stretched more than it wants to be and the restoring force acts to pull the mass back towards the equilibrium position at B. The mass of the spring-mass system, therefore, continues in simple harmonic motion by transforming its energy back and forth between the potential and kinetic forms (as is the case with the pendulum). This transformation of energy is sustained by the operation of two phenomena—Newton's first law of motion (inertia tendency) and the restoring force of the system.

One other property of the spring-mass system that we must consider is the resistance in the system. The resistance of a *S-M* system arises from the friction within the system. The friction in an *S-M* system develops internally within the spring itself and by the action of the *S-M* mass rubbing against air and support surface particles. In both cases, the system's energy is converted into heat—dissipative energy. Thus, as with the pendulum, the resistance in the spring-mass system results in damping of the amplitude of the system's motion by heat generation.

We have seen that the forces, energies and properties which control simple harmonic motion in the spring-mass system are the same as the ones that control motion in the pendulum, although they arise from somewhat different physical elements in some cases. The particular usefulness of the spring-mass system in acoustics is apparent in the study of sound wave phenomena. Most systems that can transmit sound waves are made up of individual particles. Our study of how these particles move in sustaining sound waves will be greatly simplified if we conceive of each particle as a tiny spring-mass system. This concept will be expanded when we consider wave phenomena.

CHAPTER 3
FORCED PERIODIC MOTION

In Chapter 2 we examined the properties of periodic motion under conditions where a system was set in motion and then allowed to operate without further outside influence. Whenever a system is allowed to move freely (like our pendulum in Chapter 2) we say the system is demonstrating *free periodic motion.* In this chapter we will examine a different set of circumstances. We will look at a system where an external driving force is applied to the system forcing the system to operate the way the driving force dictates. This may or may not be the manner in which the system naturally operates. Whenever a system is operated such that it is driven periodically by some outside source, we say that the system undergoes or demonstrates *forced periodic motion.* In order to study forced periodic motion, it will be useful for us to find a model that demonstrates the properties we wish to examine. The model of forced periodic motion that we will use in this chapter is the old backyard swing. A quick consideration of Figure 3.1 will reveal that the swing demonstrates the same basic properties as the pendulum we used as a model in Chapter 2. The one major difference between the swing and the pendulum is a feature upon which we will concentrate for the remainder of this chapter. The weight at the end of the arm on a pendulum is an inanimate mass that is acted upon by gravity and demonstrates inertia. By contrast, the weight on a swing is a person who is capable of exerting energy and thus adding an extra force to the pendulum-type activity of the swing. Without discussing the details of how this force is actually applied, we will simply state that by "pumping" the swing, the rider can add a *periodic driving force* to the inertia and restoring forces acting on the swing. By using your experience with the operation of a backyard swing, we will now be able to consider several basic aspects of forced periodic motion.

When you sit on a swing and "pump" it, you notice that there is a certain rate of pumping (frequency) that gives the most amount of swinging for a given amount of pumping. If you pump the swing too fast (at too high a frequency) or too slowly (at too low a frequency), the amplitude of the swinging never gets very high. If, however, you pump the swing at just the right rate (at the swing's natural frequency)

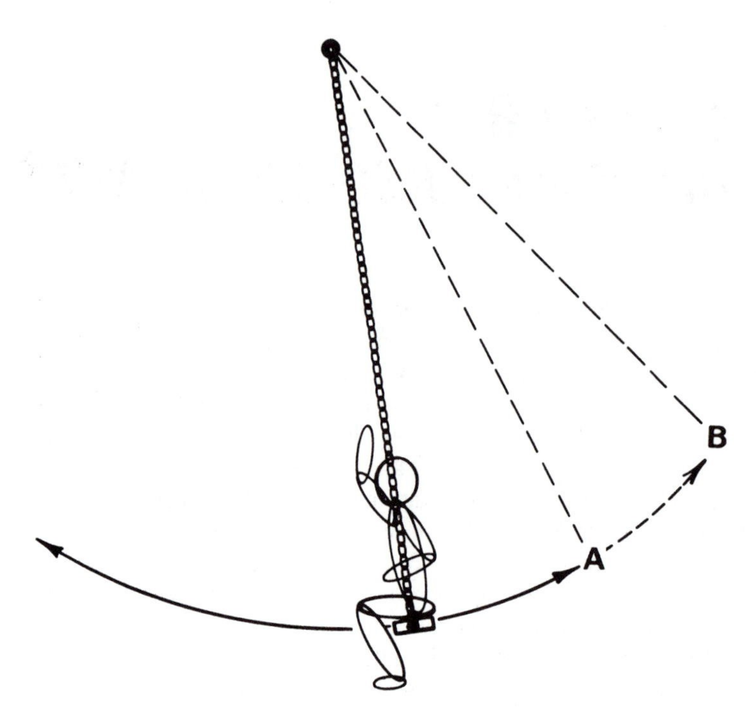

Figure 3.1. The swing: A model for forced periodic motion showing the effect of a driving force applied at varying frequencies.

you can get it to swing very high (with large amplitude) using relatively little exertion. This phenomenon of having a *natural frequency* is demonstrated by all systems that show periodic motion.

Let us consider for a while why a system might show this natural frequency phenomenon. Remember the pendulum we looked at in Chapter 2. It had two phenomena acting within it—a restoring force and inertia derived from the mass acting in compliance with Newton's first law. When the pendulum was swung, these phemonena interacted so that it would remain in motion until resistance (friction force) had dissipated all of the energy originally imparted to it. When such a system has an external driving force applied to it (like the person on the swing pumping it), this external force must be included in our consideration of the motion of the swing. In all cases that we will discuss, the push applied by the periodic driving force is the same—the swing is pushed in one direction (e.g., forward) for a certain amount of time and then the swing is pushed in the other direction (backward) for a similar amount of time. The amount of time required to go through one cycle of pumping (forward and backward) is the period of the driving force. Conversely, the number of pumping cycles (forward and backward) that can be completed in a unit of time is the frequency of the driving force.

The first case we will examine is the one in which the frequency of the driving force (or the pumping frequency) is the same as the natural frequency of the swing. Pumping is usually accomplished by the person on the swing shifting the center of gravity so that an unbalanced force is created, causing the swing to move. The person on the swing times the application of this "pumping force" to coincide with the direction of the restoring force. Thus, the two forces add together in both the forward and backward movement so that they *reinforce* each other. Since the rate of change of the driving force matches the natural frequency of the swing, the pumping changes direction at the same instant that the restoring force overbalances the inertia to change the swing's direction. Thus, when the periodic driving force on a swing is matched in timing to the natural operation of the swing all the phenomena tend to reinforce one another and the swing shows a high amplitude swinging. The use of a driving force in this way tends to replace energy in the swing that was dissipated by resistance forces. Thus, a pumped swing will not demonstrate the phenomenon of gradually reducing in amplitude due to damping. The driving force will counteract resistance so that the swing can go on moving indefinitely.

Now let us consider what will happen to a swing where the pumping frequency differs from the natural frequency of the swing. First, we will examine a case where the pumping frequency is higher than the natural frequency of the swing. In this case the swing will begin to move in one particular direction (e.g., backwards) as in Figure 3.1. When the pumping force changes direction and begins to move the swing forward again, the swing has only reached point A of Figure 3.1. The physics of the swing at its natural frequency, however, dictate that inertia will try to move the swing further backwards to point B before changing direction. Thus, the driving force must work against inertia instead of with it to move the swing forward. This same phenomenon will occur at the forward extremity of the swing's movement. The driving force will move the swing backwards but inertia will try to move the swing further forward so that the driving force is again working against inertia. It may be seen, therefore, that for much of the time during which the swing is in motion, the driving force is working against inertia. Recall that when the driving force was at the natural frequency of the swing only enough energy was needed to overcome the system's resistance. In this case of a high frequency driving force, the energy expended must overcome both resistance and inertia. Thus, at these high frequencies, it takes more energy to move the swing than at its natural frequency. Since the increase in energy required to move the swing is dependent upon the swing's inertia, we say that the system is *inertia controlled* at frequencies higher than the natural frequency of the system.

Another case where the driving force frequency of a system may be different from the system's natural frequency is when the driving force frequency is lower than the natural frequency of the system. Let us look again at our backyard swing under these conditions. For this discussion, we will refer again to Figure 3.1. This time the driving force frequency is lower than the natural frequency of the system so that the swing is moving slower than it would at its natural frequency. Because it is moving slower than usual, it only reaches backwards as far as point A of Figure 3.1 when the restoring force overcomes inertia and tries to pull the swing forward. The driving force is at a low frequency, however, and it continues to pump the swing backwards to point B of Figure 3.1. Since the restoring force is pulling the swing forward as the driving force moves it backwards from point A to point B, the driving force must work against the restoring force. A similar problem is encountered when the swing is moving forward. Thus, when the driving force frequency of the system is below the natural frequency of the system, the driving force must work against both the resistance and the restoring force of the system. This means that moving the system below its natural frequency requires more energy than moving it at its natural frequency. (Recall that at the natural frequency the driving force works against only the system's resistance.) Since the increase in force required to move the system at frequencies below the natural frequency is dependent upon the restoring force, we say that a system moving at a frequency below its natural frequency is *restoring force controlled.* Finally, since the driving force of a system moving at its natural frequency must overcome only the system's resistance, we say that a system moving at its natural frequency is *resistance controlled.*

Note that the system may be moved with the least expenditure of energy by moving it at its natural frequency. This is an important concept in acoustics. When a system is described, it is useful to be able to say how much force is required to produce a certain amount of movement at many different driving force frequencies. Knowing this is so important that a special quantity was invented to describe the force-to-movement relationship. As we noted in Chapter 2, whenever we wish to describe motion, we must select one of the properties of motion to describe it—displacement, velocity or acceleration. When we describe the relationship of force and motion in acoustics, we usually refer to the velocity component of motion. We could as easily use displacement or acceleration, but conventionally the property examined is velocity. The quantity developed, then, is the ratio of force put in (called *input force*) to the system relative to the velocity that comes out (called *output force*) of the system. This quantity is called the *characteristic impedance magnitude* of a system. It is defined mathematically by the expression

$$|Z| = F/v \tag{3.1}$$

where

F = input force, and
v = output velocity.

If you will recall what we have just said about the swing, you will see that the impedance of a system changes (varies) with the frequency of the driving force applied to the system. Recall that when the swing was pumped at the same frequency as its natural frequency, the driving force needed to overcome only the system's resistance to keep it in motion so that a relatively small force resulted in a significant velocity of system motion. At the natural frequency of the system, therefore, the F is small and the v is large so the impedance magnitude ($|Z| = F/v$) is small. When the driving frequency of the swing is higher than its natural frequency, the pumping force must overcome both the resistance and the inertia of the swing. Thus, at frequencies higher than the natural frequency of the system, the F must be larger (than it was at the natural frequency) to obtain a v the same size as the one at the natural frequency. At frequencies higher than the natural frequency, therefore, the impedance magnitude ($|Z| = F/v$) must be much larger than it is at the natural frequency. Similarly, at driving force frequencies lower than the natural frequency of the system, the driving force must overcome both resistance and restoring force so that at these frequencies F is also large. Because F is large at these frequencies, then $|Z|$ is also large at these lower than natural frequencies for the system. We may graph the magnitude of the impedance ($|Z|$) as it varies with relation to the frequency of the driving force. A graph of this type is shown in Figure 3.2. Notice that the impedance magnitude of the system is large for frequencies higher and lower than the natural frequency of the system and that $|Z|$ is relatively small at the natural frequency of the system. This general impedance function is seen in all the acoustical systems that we will study. It is an important concept in the physics of sound.

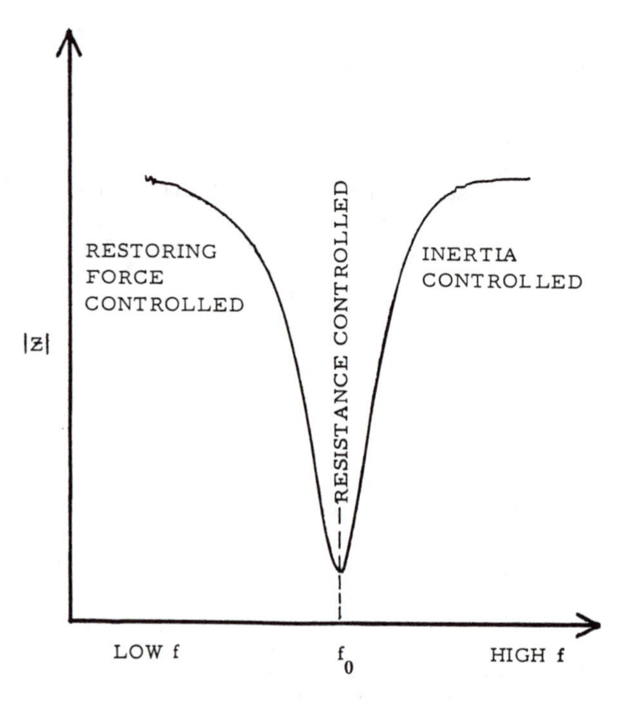

Figure 3.2. A graph of the impedance magnitude ($|Z|$) of a system as a function of the frequency of the system's driving force.

Recalling that $|Z| = F/v$ it may be seen by examining Figure 3.2. that much less force is required to obtain a given velocity of system movement at its natural frequency than at any other driving force frequency for the system. This natural frequency of periodic movement for a system is called the *resonance frequency* of the system. In fact, the resonance frequency of a system may be defined as the frequency at which the least force is necessary to achieve a given velocity of motion. That is, the resonance frequency of a system is the one at which the system's characteristic impedance is at its smallest. The physicist would say, "At the resonance frequency of a system, its characteristic impedance is a *minimum*."

At this point, we will consider another relationship inherent in the periodic motion of a *driven system* such as our pumped swing. Recall the description of the motion of the swing when it was being pumped at a frequency below its resonance (natural) frequency. The system was forced to move at a frequency lower than that at which inertia and restoring force were trying to move it. It is often useful in acoustics to discuss the combined effect of the two phenomena called inertia and restoring force. The combined effect of these two phenomena (inertia and restoring force) is, therefore, described by a single term—the *reactance* of the system. In effect, at the lower than resonance frequency, the swing is moving slower than the reactance wants it to. Another way to say this is that the time pattern of the swing's movement lags behind (or is slower than) the time pattern of the reactance. The time relationship between two phenomena is referred to as the *phase relationship* of the phenomena. The phase relationship is commonly reported in degrees, as of an angle. Thus, the phase difference between the actual movement of the system and the way in which the reactance wants the system to move is called the *phase angle of the system impedance* (ϕ) and this quantity also varies with frequency. At frequencies below resonance, the system movement lags behind the reactance-dictated movement so the impedance phase angle is negative (less than zero). By contrast, at frequencies higher than resonance, the system movement leads the dictated movement of the reactance in time and the impedance phase angle is positive (larger than zero). Finally, when the system is moving at its resonance frequency, there is no difference between the actual system movement and the movement dictated by the reactance, so the impedance phase angle is zero. It is also possible to graph the value of the impedance phase angle as it varies with frequency. This is done in Figure 3.3. The graph of impedance magnitude is also shown again on this graph. Note that for frequencies below resonance the phase angle is negative and for frequencies above resonance the phase angle is positive. Also observe that the resonance frequency could be alternatively defined as the frequency at which the impedance phase angle is zero. A final point to note about these two graphs is that they are both necessary to completely define the impedance of the system. If we know only the magnitude of an impedance we know the force-velocity ratio of the system for a given set of conditions but we cannot tell whether we are above or below the system's resonance frequency. This occurs because (for the frequencies of interest) the system's impedance magnitude function below resonance is a mirror image of the function above resonance (see Figure 3.3). Furthermore, if we were to know only the impedance phase angle of a system for a given set of conditions, we would know if we were below or above resonance but we would be unaware of the force-velocity ratio for the system under those conditions. Thus, in specifying the operational characteristics of a system, it is necessary to define the way in which both the impedance magnitude and the impedance phase angle vary with frequency. This is

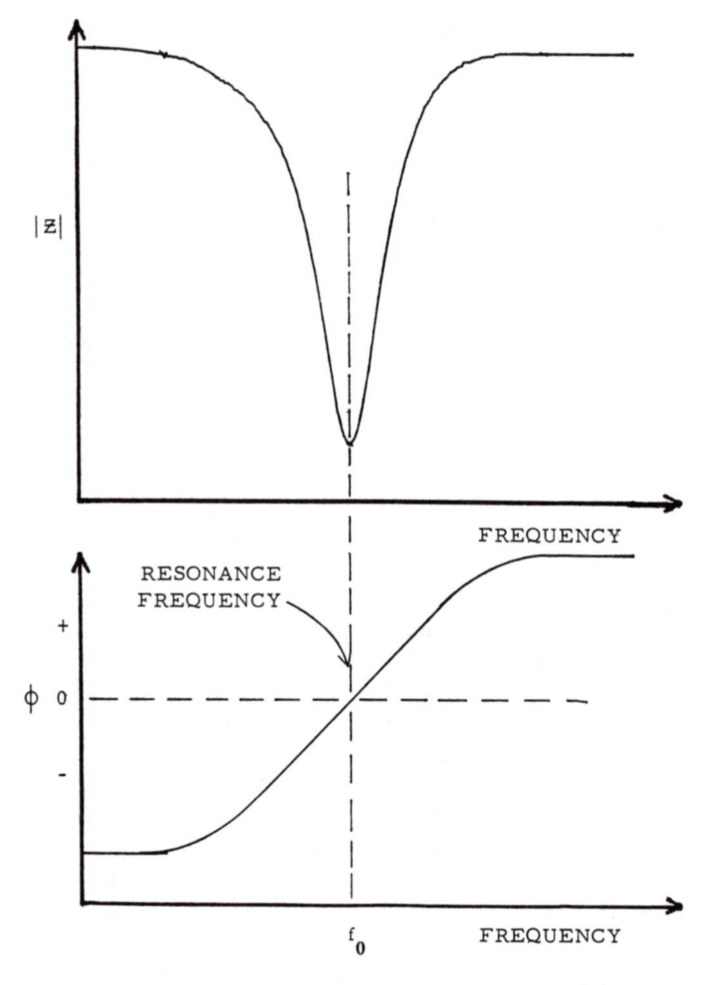

Figure 3.3. Graphs of impedance magnitude ($|Z|$) and phase angle (ϕ) as functions of frequency.

often done by presenting the two graphs for a given system as shown in Figure 3.3.

The impedance function is not the only way in which the operational features of a system may be described. We will now consider an alternate way of describing the motion of our swing. Suppose we find that the person riding the swing is able to produce the *exact same pumping force at all driving frequencies*. Since we know that the impedance magnitude varies with frequency and since $|Z| = F/v$, then if F is constant at different frequencies, v will vary with frequency. At the resonance frequency (usually designated as f_o), the impedance magnitude is the smallest, so (because impedance magnitude is *inversely* proportional to velocity) velocity must be its largest. As the system's frequency of motion gets farther away from the resonance frequency in either the high or low frequency direction, the impedance magnitude grows larger. Thus, the velocity output of the system for a given amount of input force grows smaller as the system's operating frequency moves away from the system's resonance frequency. A graph of the system's velocity output, showing how it varies with frequency, is given in Figure 3.4. This type of graph is called the *frequency response* of a system. The frequency response of a system may be plotted in output velocity (as we did), output displacement, output acceleration or some other useful measure of a system's motion. Different parameters are used in different cases because in many cases we are interested in one particular aspect of a system's motion. It is important to know what parameter of motion is being plotted in a given frequency response chart.

In studying a frequency response of a given system, we wish to examine certain specific features. One of the most interesting aspects of a system's frequency response is called its quality factor and is

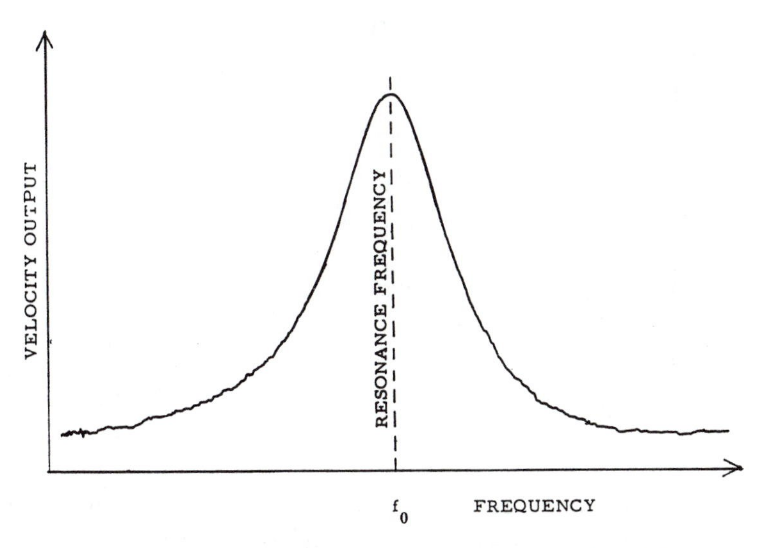

Figure 3.4. The frequency response of our swing. Note: The pumping force is assumed to be the same at all frequencies.

symbolized by the letter Q. The Q of the system denotes how sharp the system's response is at the resonance. In other words, Q designates how much higher a system's resonance frequency response is compared to its response at non-resonance frequencies. If Q is high, the peak in the frequency response is high; whereas, if Q is small, the frequency response peak at f_0 is shallow. This phenomenon is shown in Figure 3.5. The systems in this figure have the same resonance frequency but they show a markedly different Q. The reason that different systems show different values of Q is dependent upon some ideas developed earlier in this chapter. Recall what we said about the forces that must be overcome by the driving force of a system. At the resonance frequency, the driving force need only overcome the mechanical resistance, R, of a system. By contrast, above the resonance frequency, the driving force must overcome both inertia and mechanical resistance; while below the resonance frequency, the driving force must overcome restoring force and mechanical resistance. Thus, at all frequencies, the driving force must overcome resistance, R, and this resistance sets an upper limit to the amount of response that can be obtained from a given amount of force. At the

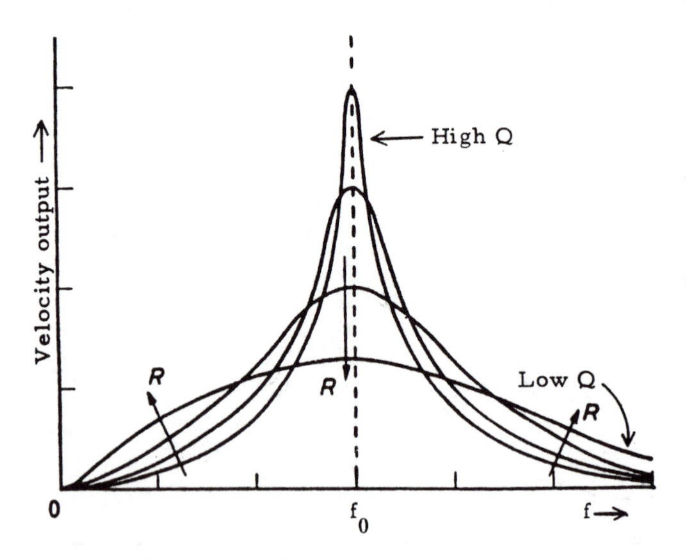

Figure 3.5. Frequency response curves for velocity amplitude with different values of mechanical resistance R. The arrows indicate how the curves change with increasing amounts of mechanical resistance.

non-resonance frequencies, however, some reactance component (either inertia or restoring force) further subtracts from the system's output for that same applied driving force. Thus, the height of the resonance peak relative to the off-resonance response of a system depends upon the difference between the magnitude of the system's mechanical resistance and the magnitude of its reactances. If a system has a small mechanical resistance component, the driving force will have little opposition at the resonance frequency and output will be high. However, at the non-resonance frequencies, the driving force faces the opposition of the small resistance and a large (by comparison with the resistance) reactance which results in a considerable drop in output for these frequencies. By contrast, consider a system where mechanical resistance is large compared to the reactance. Here we see a situation where output will be relatively small at resonance because the driving force is dissipated by a large mechanical resistance. In this case, at non-resonance frequencies, the force opposing the driving force is increased by a small amount (compared to the mechanical resistance). Thus, the output at non-resonance frequencies is less than at resonance but only by a small amount. In summary, we may say that systems with small amounts of internal mechanical resistance (low damping) demonstrate a *high-Q operation*. Those systems with large internal mechanical resistances, however, (high damping) demonstrate *low-Q operation*.

Up to this point, we have considered the forced periodic motion of a system with one simple outside driving force. What would happen to our swing if two pushing forces were applied to it? How would these pushing forces interact? We will consider briefly in the next few pages exactly what does happen when the driving force applied to the system becomes more complex. We will begin by looking again at our swing when it is pumped at its resonant frequency by the person riding on it.

Let us look at a time/amplitude plot of the velocity of the swing's movement. The person riding the swing is pumping just enough to overcome the swing's mechanical resistance and keep it moving in a uniform manner. The time/amplitude plot of this motion is shown in Figure 3.6. Now we will add a second driving force to the swing system. Suppose a person were to stand behind the swing and give it a push on every second cycle. The person riding the swing would continue pumping exactly as he had before. In such a case, the swing would have an extra driving force pushing in the same direction as the pumping. Since they are in the same direction, these forces would add, thereby changing the swing's motion. Because the second pushing force is applied only on every second cycle, the swing's mechanical resistance force would dissipate the energy from this second force during the two intervening

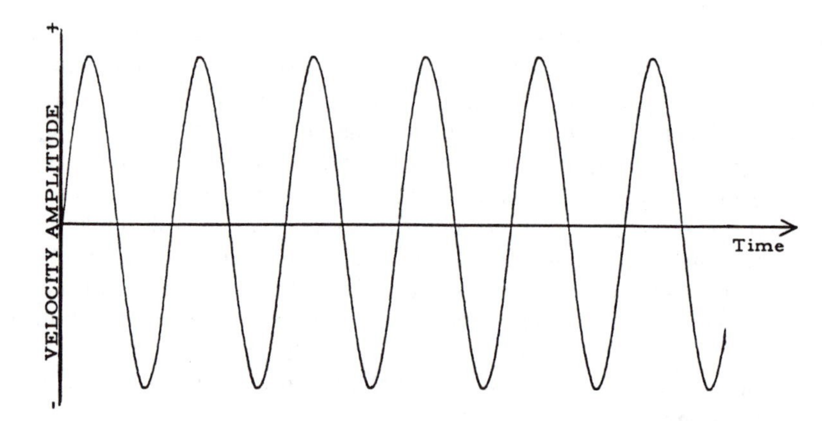

Figure 3.6. Amplitude versus time plot of the movement of a swing "pumped" at its resonance frequency with just enough force to overcome resistance.

cycles. This type of motion is shown in Figure 3.7. Note that the movement of the swing is still periodic but that the amplitude of motion varies. The second pushing force is applied on every second cycle of the swing's motion so the frequency of the second force is one half of the pumping force. Note that while the swing is moving at the frequency of the pumping force, its amplitude is varying at one half that rate (i.e., the amplitude is varying at the frequency of the second or pushing force). Thus, the various components of different driving forces are represented in the more complex movement of the swing. It is in fact possible to examine the movement of any body in periodic motion and, by analyzing that motion, to determine the frequencies of all the component motions giving rise to the body's movement. In many cases, the movement of systems may include a very large number of component motions and the *analysis* process can become very complex. This process of determining the simple, component motions making up a more complex motion is called *Fourier analysis*. It is named after the French mathematician who developed the method for performing this complex analysis.

Thus, we have seen that two forces acting in the same direction simply add to each other when they are both applied to the same system. In fact, this addition rule precisely defines how forces interact. If the two forces are of the same magnitude, the system's amplitude of motion will be doubled since the magnitude of the applied force is doubled. If the magnitude of the second force is twice that of the original force, the total force applied to the system is tripled and the amplitude of the resultant system motion will also be tripled. If the forces are acting in

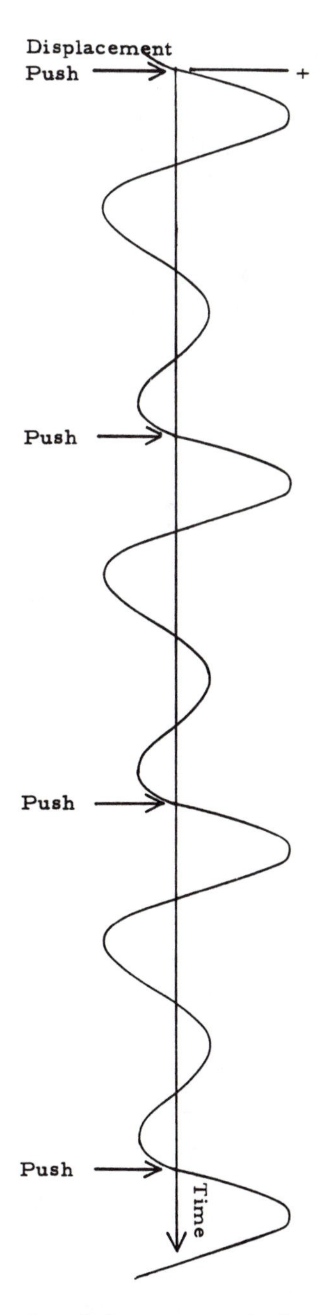

Figure 3.7. Amplitude plot of the movement of a swing with two applied driving forces. Swing movement is maintained by "pumping" action of person on the swing while a second driving force is applied every second cycle by a person on the ground.

opposite directions, then they will subtract from one another in the same manner that they added before and the resultant system motion will decrease in magnitude. In fact, if two opposing forces of equal magnitude are applied to the system no motion will result. This can be seen by considering the swing again. If the rider of the swing is pumping it backwards with a certain force while the standing person is pushing the swing forward with an equal force, the swing will not move. This

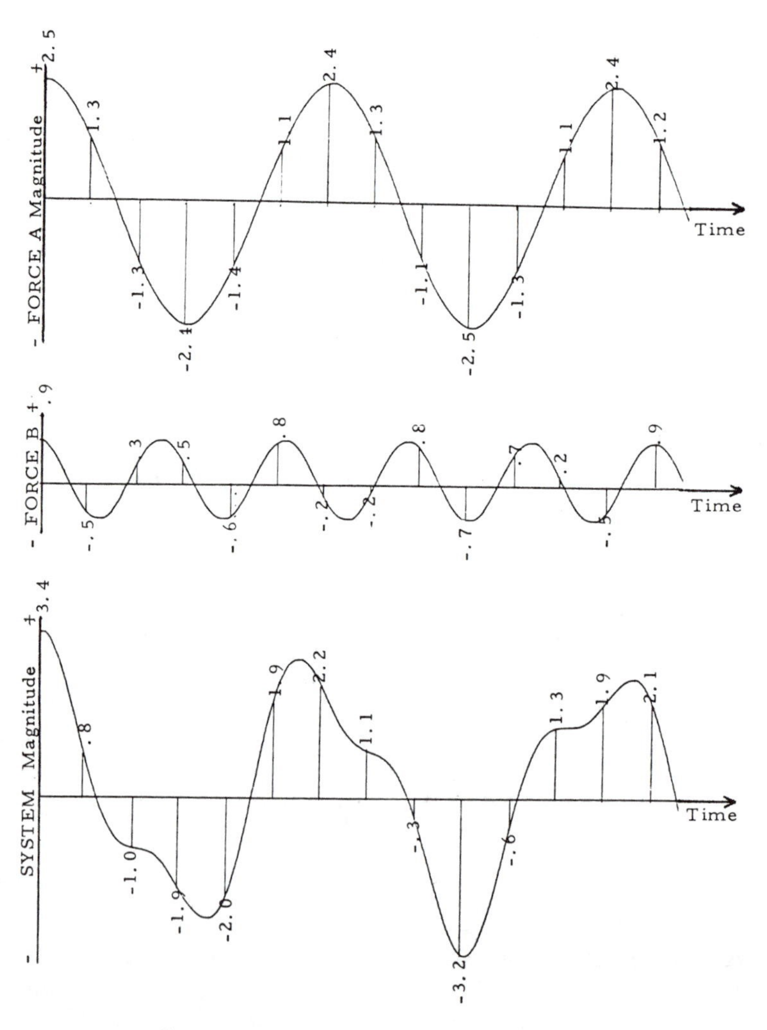

Figure 3.8. Addition of periodic forces.

phenomenon is called *cancellation.* In general, we may state that whenever two periodic forces are applied to a system, the resultant system motion will be proportional to the arithmetic sum of the two forces. This addition principle is shown in Figure 3.8. The proportional system amplitude at the bottom of the figure is the arithmetic sum of the amplitudes of the two driving forces. Note that the resultant motion indicated by the lower curve is a distorted version of the motion shown in the uppermost curve. If these represented pressure changes associated with audible frequencies, the sound heard would be described as "harsh" or "buzzing". If the two frequencies being combined were very nearly the same, the resulting sound would gradually increase and decrease in loudness. This rising and falling in loudness is called *beating* and we say the two tones together produce *beats.*

Up to this point, we have considered forced periodic motion only in an ongoing condition—after being started up and before slowing down. Ongoing motion of this sort is called *steady state activity.* In reality, when an external force begins to act upon a system, it takes

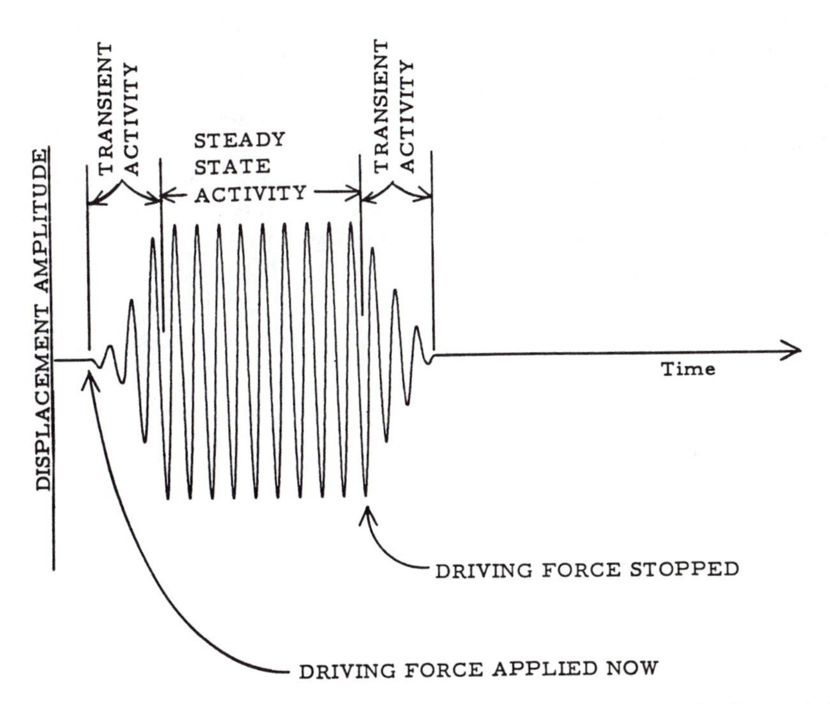

Figure 3.9. Amplitude versus time plot of a system's displacement showing both steady state and transient activity.

some time for the buildup of motion to reach a steady state condition. When a system is in the process of building up to its steady state condition, we say the system is demonstrating *transient activity*. Transient activity is especially difficult to study because it is a state of constantly changing motion. It is of importance in the study of the speech and hearing sciences because much of the activity involved in the production and reception of speech is transient in nature. Transient activity is also demonstrated by a system when its driving force is removed. Once the driving force stops pushing a system, it takes some time for the system to stop moving. The amplitude of the system's motion gradually decreases due to damping (as described in Chapter 2) and the system eventually comes to rest. Figure 3.9 shows a system's displacement, time/amplitude plot including the transient activity at the beginning and end of the system's motion.

This chapter examines forced periodic motion. While it is easier to study motion that is periodic, we must recognize that motions can and do repeat themselves at irregular intervals; thus, not being periodic. A system in motion may constantly change its frequency of operation as it is moving if it is a driven system. When a motion is not periodic, it is called *aperiodic motion* or *noise*. Noise is an important type of motion for the study of the speech and hearing sciences because the speech signal of such vital concern to us in this science is aperiodic (i.e., noise). While we will continue to study systems operating at a single frequency, it is important for the student of the speech and hearing sciences to keep in mind that speech is aperiodic and is thus a more complex case than many simple ones we will examine in this text.

CHAPTER 4
WAVE PHENOMENA

In the preceding chapters we have established the principles of linear and periodic motion. Having reviewed this material, we are now ready to consider those phenomena which are more readily recognized as falling within the specific realm we call acoustics. We will concern ourselves in this chapter with acoustic waves.

The concept of wave motion is encountered in many areas of physics. When we speak of a wave, we are speaking of an energy transportation mechanism. We may say that a wave travels from point A to point B. When this has happened, some form of energy is transported from point A to point B but (and this is a critical factor) matter is not transported between the two points—only some form of energy. This is not an easy concept to grasp and some amount of elaboration is necessary for its understanding. We will consider a specific example of wave motion: ripples on the surface of a pond.

Visualize, for a moment, a small pond surrounded by a forest. Everything is quiet and calm. There is no hint of a breeze and the water surface is as smooth as a freshly waxed floor. On the far bank a small boy is seated among the rushes. Suddenly he takes a small stone and lofts it high into the air so that it plunges smoothly into the water with hardly a splash. Immediately ripples radiate out from the point of impact in ever enlarging rings. Near their center, the rings are clearly defined and easy to see, but as they spread farther from their source they become increasingly indistinct until they gradually fade out altogether.

This scene contains all of the elements necessary for a discussion of the fundamentals of wave phenomena. Initially, the water in the pond is standing still. The water is the material within which the ripples (waves) will travel. The material which sustains wave motion is referred to as the *medium*. The medium is made up of small particles of matter (in this case, particles of water). When we say that the medium is essentially motionless, we do not mean that the particles of the medium are all standing still. The particles of any medium will be moving randomly under normal conditions, but since their motion is random there is no total observed motion of the medium and we say that it is still. By contrast, when a medium is set into observable motion, the particles of that medium are made to move systematically in the same manner rather than randomly. The total systematic movement of the particles results in observable movement of the medium. Thus, initially the particles of water in the pond were moving randomly and only slightly, so that an observer on the shore would say that the pond was "mirror still".

At a particular instant in time, the stone thrown by the boy plunged into the water and greatly disturbed that medium. As the stone moved downwards it pushed and pulled water particles down with it, thus displacing some of the particles of the medium. There is a force in the water, however, that operates to make the particles return to their original position. Without discussing this force in detail, we may simply state that there is a *restoring force* in the water that pulls the water particles back to their starting position. We have noted before, though, that all matter complies with Newton's first law of motion—the inertia tendency. It is this inertia tendency that will make the upward moving water particles overshoot or move past their original position. The restoring force will, then, begin pulling the water particles downward in an attempt to return them to their original or *equilibrium position*. This up and down particle movement with a tradeoff occurring between the actions of the restoring force and the inertia should sound quite familiar by now. In effect, the motion of the particles of water is the same type of motion we examined in our study of the pendulum in Chapter 2. The water particles have entered into periodic motion. They have, in fact, begun to move in the manner we have described as *simple harmonic motion*. As a matter of fact, the general rules of motion which we developed in both Chapters 2 and 3 apply to the motion of the particles of water we now wish to examine.

PROPAGATION

Up to this point, we have described nothing which has not been previously examined under slightly different conditions. Now we will introduce a new concept—the phenomenon of *wave propagation*. Water displays a property referred to as contiguity. A *contiguous medium* is one in which the particles of the medium are in close proximity to one another. Almost all matter displays this property of contiguity to some degree or other. The particles of contiguous media are, in fact, in close enough proximity to one another that they tend to attract one another. Because of this property, it is impossible for one particle of such a medium to enter into simple harmonic motion without affecting its neighboring particles. When, for example, the water particles struck by the stone enter into simple harmonic motion, they tend to drag along adjacent particles due to their attraction for one another. Thus, the stone will set a few water particles in motion. These particles will set adjacent particles in motion. Those particles will transfer the motion to still other particles and so on until a large area of the medium, in this case the water, has been set in motion. This is the phenomenon that gives rise to the ripples radiating outward from the point where the stone originally struck the water surface.

There is a critical point about this phenomenon that must be clearly understood. *Wave propagation* is a means of transferring energy from one place to another but no matter is transported. Figure 4.1 has been developed to assist the student in understanding this fact. The figure shows a series of sketches of a set of seven particles of a contiguous medium. The set of sketches represents the particle movement of each of the seven particles at consecutive instants in time. Initially all particles are at rest (moment A). An instant later, at moment B, particle 1 is moving downwards, as shown by the arrow, while the other particles are essentially motionless. An instant later (moment C), particle 1 is moving upwards past its equilibrium position and some motional energy has been transferred to particle 2 which is now moving downwards. This process continues with the motional energy being transferred to particles farther and farther to the right. In moment D, particle 3 has begun to move; at F, particle 5.

As the motional energy is transported to the right, the particles that were originally in motion gradually return to equilibrium as they have expended their motional energy in exciting their neighbors. Thus, at moment *F*, particle 1 is again at equilibrium. At moment *G*, both particles 1 and 2 are at equilibrium. Note that while motional energy has traveled from left to right (beginning at particle 1 and reaching particle 7), the motion of any given particle is entirely up and down. Thus, while the wave propagation resulted in a transportation of motional energy from left to right, there was no movement of the particles of the medium in that direction. This fact is a unique feature of wave phenomena and must be kept in mind whenever such occurrences are studied.

The motion we have described this far is the type we see on the surface of a pond whenever it is disturbed. The particles move up and down as the wave propagates from left to right. This type of wave motion is referred to as transverse wave motion. In *transverse wave motion*, the particles of the medium move at right angles to the axis

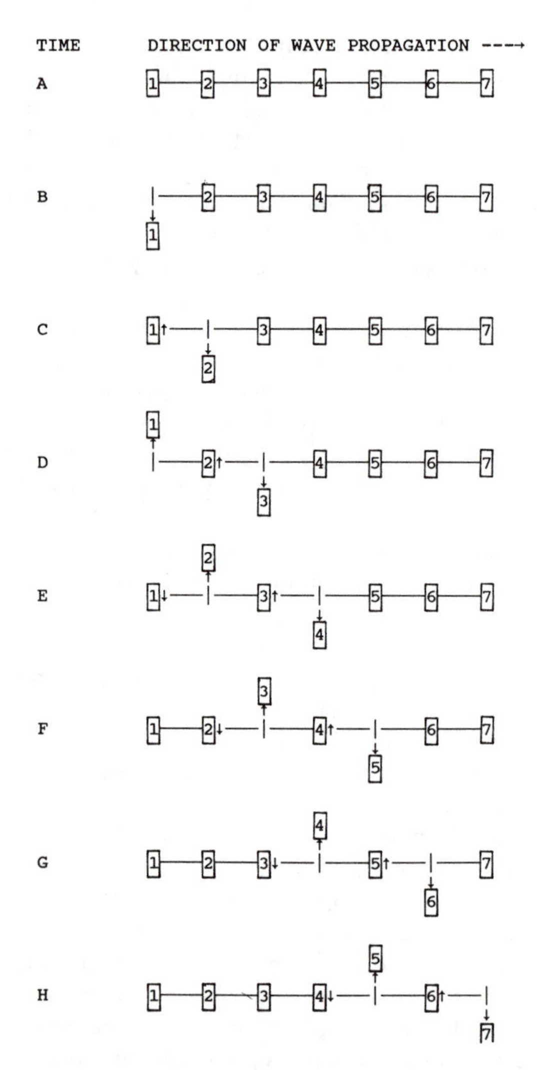

Figure 4.1 Propagation of energy but not of matter

along which the wave is propagating. There are, however, other types of wave motion as depicted in Figure 4.2. In *longitudinal wave motion*, the motion energy makes the particles move back and forth in a direction parallel to the axis of propagation. While in this case particles move a bit from left to right, the restoring force still returns them to the

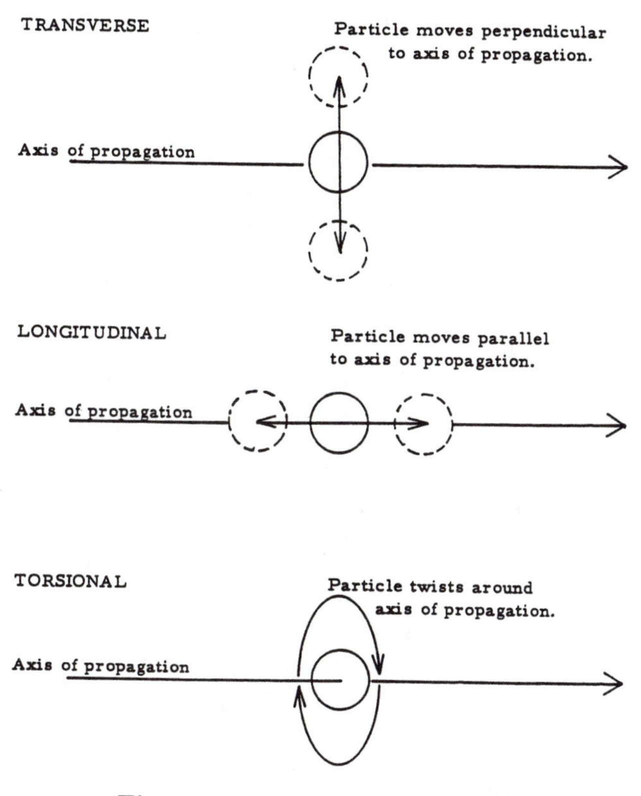

Figure 4.2. Types of wave motion.

equilibrium position so that only energy and not matter is transported by the wave propagation. After the passage of a longitudinal wave, the particles of the medium are distributed in about the same positions as they were before the passage of the wave. The longitudinal wave is of particular interest because sound waves in air demonstrate the longitudinal type of wave motion. A final type of wave motion is torsional. In *torsional wave motion*, the particles of the medium twist around the axis of wave propagation. This is one manner in which sound waves travel in bars of solid matter such as the bars of a xylophone.

In the next part of this chapter, we will consider sound waves traveling in air. Thus, we will concern ourselves with longitudinal waves. When any sound source radiates what we call a sound, what actually happens is that it pushes on the air particles adjacent to it. For an instant, then, some air particles move in the same direction as the wave propagation. This movement causes the group of air particles to get bunched together for an instant. We say that the *compression cycle* of the wave has reached that point at that instant in time. As time passes, the original particles are pulled back where they came from by a restoring force. In addition, other particles, pushed by the crowding in of the original particles, move forward along the axis of propagation. This leaves a relatively empty space where an instant ago there was a compression. Such a relatively empty condition is referred to as the *rarefaction cycle* of the wave. Thus, a sound source is constantly radiating waves of compression, then rarefaction, then compression, etc., into the air around it. When these waves are received by our hearing mechanism, we perceive what we call a sound. The distance between two consecutive compression peaks in space is called the *wavelength* of the sound and it can be measured in feet or meters. Wavelength is generally symbolized by the Greek letter lambda, λ.

Let us consider for a moment a small spherical sound source radiating sounds into the air in all directions. We will say that the *sound power* of the source is a measure of the amount of sound the sound source can produce. The symbol for sound power is W since power is measured in watts. Let us further stipulate that this particular sound source has a constant sound power output (W) so that a given and constant amount of sound power is imparted to each sound wave that radiates from the source. In the absence of any obstructions, these sound waves will radiate uniformly in all directions as ever-enlarging, spherical wavefronts of sound. This unobstructed radiation of sound is often referred to as *spherical divergence* of sound and the waves themselves are called acoustic *spherical waves*. The acoustic spherical wave has only as much sound power as it originally got from the source but, since the wavefront is an ever-increasing sphere, the amount of sound power at any point on the wavefront must be continuously decreasing as its distance from the source increases. Think of the sound wave as a balloon and the sound power as the amount of balloon material available. The more the balloon is inflated, the larger it gets and the thinner the balloon's skin. The balloon material must be spread thinner and thinner so as to cover the ever-inflating balloon. Similarly, the sound power in the sound wave must be spread thinner and thinner to cover the ever-enlarging wavefront. *Sound intensity* is defined as the amount of sound power in a given area of the sound wavefront. Recall from geometry the formula for the area of the surface of a sphere:

$$A = 4\pi r^2 , \tag{4.1}$$

where r is the radius of the sphere. Since intensity is power per unit area and the area of the wavefront is a sphere, the formula for sound intensity must be:

$$I = \frac{W}{4\pi r^2}\,. \tag{4.2}$$

Note that if we ignore parts of this equation for a moment, we can see that intensity is proportional to $1/r^2$:

$$I \propto 1/r^2\,.$$

This is not an equation. The symbol "$\propto$" means "is proportional to" and is used to show a proportionality—an important part of a larger and more complex equation. The proportionality we have selected says that the intensity of a sound wave is inversely proportional to (i.e., is proportional to 1 over) the square of the distance from the source. This is known as the *inverse square law* and is an important relationship in acoustics. It shows that the intensity of a sound decreases very rapidly as we move away from the source of that sound.

Unfortunately, there are no devices that respond directly to sound intensity. Microphones and ears respond to the pressure produced by sound waves. *Sound pressure* (p) is the amount of force (i.e., push) in a given area of the sound wavefront. The unit of pressure is the *Pascal* which has dimensions of Newtons per square meter. The microphone or ear then simply responds to how hard the sound wave pushes on some receiving surface such as an eardrum or microphone diaphragm. There is a direct relationship between sound intensity and sound pressure that depends upon the characteristic impedance (ρc) of the medium that transports the sound. Intensity then is also defined by the equation:

$$I = \frac{p^2}{\rho c}\,. \tag{4.3}$$

The most significant aspect of the concept of sound intensity is that it allows us to determine the relationship between sound power and sound pressure. Recall that:

$$I = \frac{W}{4\pi r^2}$$

and that

$$I = \frac{p^2}{\rho c}\,.$$

Therefore, we may write that

$$\frac{W}{4\pi r^2} = I = \frac{p^2}{\rho c},$$

or

$$\frac{W}{4\pi r^2} = \frac{p^2}{\rho c}. \tag{4.4}$$

Thus, if we know the acoustic power produced by a source, we can determine the sound pressure it would cause at a given point some known distance from the source. Similarly, by knowing the sound pressure produced by a source at some known distance, we can calculate the sound power of the source.

A useful concept for the study of certain wave phenomena is that of the acoustic plane wave. Just as the surface of the Earth (a sphere) seems flat to us who are so small in comparison to the Earth, so too with sound waves. If we study a small part of an acoustic wavefront and we are quite far from the sound source (at least 6 or 7 wavelengths from it), the wavefront of an approaching sound wave is practically a flat surface. Such an uncurved wavefront is referred to as an *acoustic plane wave*. Technically, we may define a plane wave as one that exhibits the same sound pressure at all points on any given plane perpendicular to the axis of propagation. (A plane is an infinitely large, flat surface.) Since the wavefront of the acoustic plane wave is flat, it provides a simpler form of interaction with obstructions and other waves than the spherical wave. The plane wave is, therefore, a particularly useful model for studying wave phenomena and we will concentrate on it for the remainder of this chapter.

BOUNDARIES

One of the principle features of wave propagation that we can study by using the plane wave model is the effect of boundaries. A *boundary* is the edge or termination where two adjacent media for sound transmission come into contact with one another. The two media may be very similar such as warm air and cold air or they may be very different from one another such as air and water or air and guitar string. Regardless of what the two media are, the area where they meet is a boundary and certain phenomena will generally occur.

Whenever a sound wave impinges upon a boundary, two sound waves are generated at the boundary (see Figure 4.3). The energy in the original sound wave (called the *incident sound wave*) is broken up. Some of the energy is not accepted by the second medium (medium B) and is pushed back into the first medium (medium A). This process is called *reflection* and the sound wave that travels back into medium A is called the *reflected sound wave*. The energy that is not reflected back into medium A is accepted by medium B. This process of accepting sound energy across a boundary is called *transmission* and the sound wave that travels into medium B is called a *transmitted sound wave*. Note that at the boundary the energy from the incident sound wave is either transmitted or reflected. Energy generally does not disappear or get used up at a boundary. Thus, the total of the energies in the transmitted and reflected sound waves is equal to the energy content of the incident sound wave.

One other interesting phenomenon which may be observed at boundaries is called refraction. When the characteristic impedances (ρc)

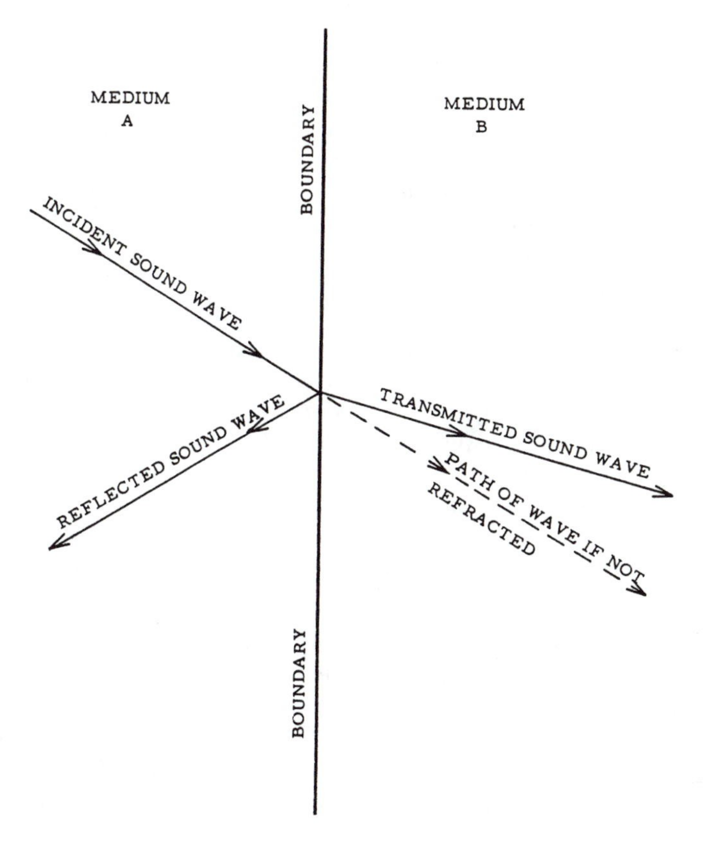

Figure 4.3. A sound wave reaching a boundary.

of two media differ from one another, the two media will differ in how sounds propagate through them. This difference causes a bending of the axis of sound propagation at the boundary. This bending of the sound propagation axis is called *refraction* and is shown in Figure 4.3. The dashed line shows the path the transmitted wave would have taken if it were not refracted. The refraction causes the transmitted propagation axis to be bent upward at the boundary. There is no refraction process in the reflected wave. In all cases, the angle between the incident wave and the boundary is equal to the angle between the reflected wave and the boundary.

The relative amplitudes of the reflected and transmitted waves depend upon two factors:

(a) the characteristic impedances (ρc) of the two media and
(b) the angle of incidence of the incident wave (the angle between the wavefront and the boundary).

In this text, to avoid discussion of complex phenomena, we will consider only the case where the wavefront is parallel to the boundary. In this case, known as *normal incidence*, the axis of propagation of the incident wave forms a right angle (90°) with the boundary. For this case, the only factor controlling the amplitudes of the transmitted and reflected waves is the characteristic impedances of the two media.

Without spending a great deal of time on mathematical derivation, we will simply state the formula for determining what proportion of the incident wave is reflected at the boundary. The primary components of the formula are the characteristic impedance of the first medium ($\rho_1 c_1$) and the characteristic impedance of the second medium ($\rho_2 c_2$) since it is a ratio of these two parameters that controls boundary phenomena. The value computed is called a *reflection coefficient* (α^r) and it is the proportion or fraction of incoming energy that is reflected at the boundary. The formula for the reflection coefficient is:

$$\alpha^r = \left[\frac{\rho_2 c_2 - \rho c_1}{\rho_2 c_2 + \rho_1 c_1}\right]^2 . \tag{4.5}$$

The range of values of α^r can theoretically go from 0.0 to 1.0. When $\alpha^r = 0.0$, the sound energy is completely accepted by the second medium and there is no reflected sound wave. When $\alpha_r = 1.0$, the sound energy is completely reflected into the first medium and there is no transmitted sound wave. In all cases, the energy content of the reflected wave (E_r) is equal to the energy content of the incident wave (E_i) multiplied by α_r. Thus:

$$E_r = \alpha_r (E_i) . \tag{4.6}$$

We may also wish to determine the energy content of the transmitted wave. To find this value, we will need to calculate a *transmission coefficient* (α_t) which is the proportion of energy transmitted into the second medium. Recall that the sum of the energies in the reflected and transmitted waves is equal to the total energy in the incident wave. Thus, all the energy that is not reflected is transmitted and the maximum possible value of a given reflection or transmission coefficient is 1.0. So we can simply compute the transmission coefficient by the equation:

$$\alpha_t = 1.0 - \alpha_r \, . \tag{4.7}$$

The energy content of the transmitted wave (E_t) would then be:

$$E_t = \alpha_t \, (E_i) \, . \tag{4.8}$$

RADIATION IMPEDANCE

Before leaving our discussion of wave phenomena, we will consider several aspects of sound radiation. The first of these is the phenomenon of radiation impedance. As with the pendulum, sound sources display a characteristic impedance that varies with frequency. Because the characteristic impedance of a sound source affects the manner in which that source radiates sound waves, it is referred to as the *radiation impedance* of the source and it is defined technically as the ratio of the force exerted by the source on the medium to the velocity of motion of the source. Recall from earlier in this chapter that the greater the similarity in the impedances of the media on two sides of a boundary, the more sound energy is transmitted across the boundary. The surface of the source may be visualized as a boundary. In order for the source to radiate sound effectively, therefore, it is desirable that the radiation impedance of the source be similar to the characteristic impedance (ρc) of the medium. In fact, it may be shown that the radiation impedance of a source becomes most similar to the ρc of the medium at frequencies where the wavelength of the sound is three times the diameter of the source. Since wavelength increases as frequency decreases, then only large sources will effectively radiate low frequency sounds. For example, a sound of 1,000 Hz has a wavelength of about 34 centimeters (~1 foot) in air, thus it is best radiated by a source 11.4 centimeters or about 4.5 inches in diameter. By contrast, a sound of 100 Hz (very low pitched) has a wavelength of about 3.4 meters (~10 feet) in air and is best radiated by a source about 1.1 meter (~3.5 feet) in diameter. It is for this reason that bass drums and bass loudspeakers must be large. This is why small loudspeakers, such as those found in a pocket radio, are really only suitable for producing high frequencies and therefore sound "tinny."

DIRECTIONALITY

A second effect of source size is directionality. *Directionality* refers to the degree to which a sound source tends to radiate sound in a particular direction. When a sound source radiates sound much more along one particular axis than in any other direction, the source is said

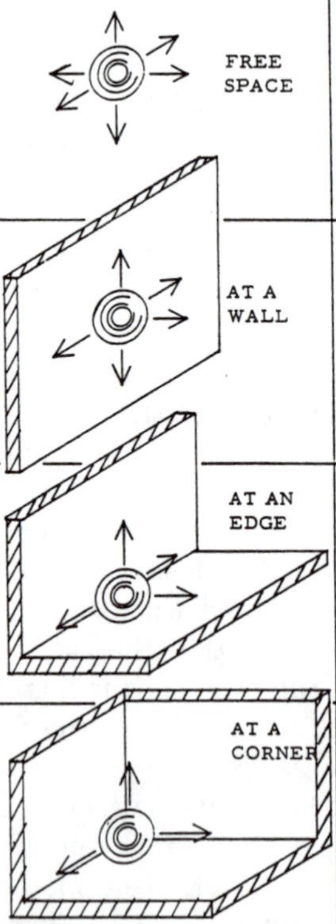

CONDITION	AREA OF WAVEFRONT	INTENSITY	PRESSURE	dB
FREE SPACE	$A = 4\pi r^2$ SPHERE	I	p	0
AT A WALL	$A = 2\pi r^2$ HEMI-SPHERE	2I	$p\sqrt{2}$	+3
AT AN EDGE	$A = \pi r^2$ QUARTER SPHERE	4I	2p	+6
AT A CORNER	$A = \frac{\pi r^2}{2}$ EIGHTH SPHERE	8I	$p\sqrt{8}$	+9

Figure 4.4. Baffle effects. dB gain is calculated relative to free-field condition.

to be *highly directional.* Alternatively, when a source radiates sound in all directions equally, we say that the sound source is *omnidirectional.* The larger the wavelength (i.e., the lower the frequency) of a radiated sound wave, the more omnidirectional it becomes. This directionality effect is closely related to the size of the sound source. For a given signal frequency, the larger the sound source, the more directional the sound wave. Keep in mind, however, that due to radiation impedance this phenomenon is limited to a small range of source sizes for any given frequency. We may, therefore, state that the degree of directionality (D) of a given source is related to two factors:

$$D \propto a/\lambda ,$$

where a is the radius of the source.

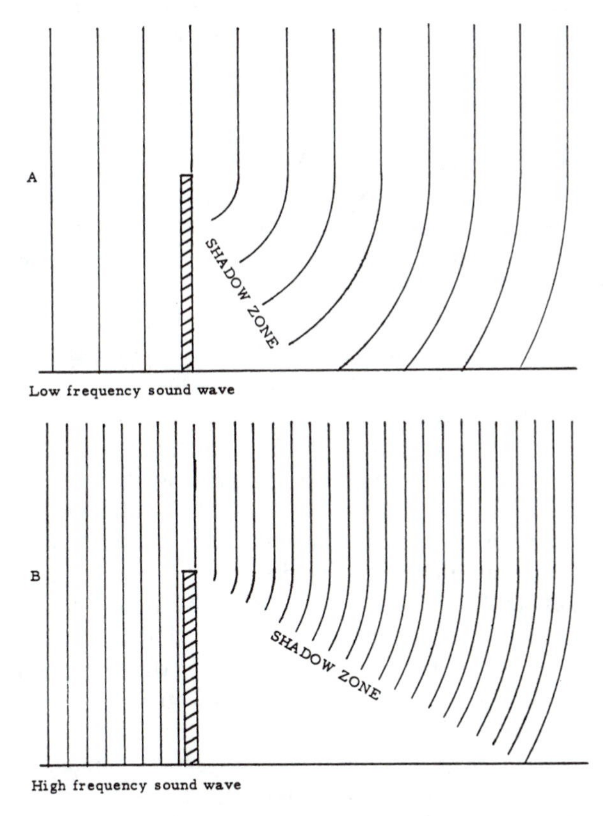

Figure 4.5. Diffraction of plane waves by a barrier.

BAFFLES AND BARRIERS

A final phenomenon that we will consider in this chapter is the effect of baffles. A *baffle* is a barrier or wall that in some way restricts the propagation and/or the radiation of sound waves. Recall that the sound pressure we can measure coming from a source in free space is an amount of pushing force exerted on the microphone diaphragm by the sound wave. The sound wave gets its pushing force from the power of the sound source and this acoustic power is spread evenly over the entire spherical sound wavefront. The area of the wavefront is given as $4\pi r^2$ (where r is the distance from the source). This situation is summarized in the top row of Figure 4.4.

Now suppose we were to situate our sound source by a large wall. In this case, the sound cannot radiate into the wall and so a spherical wavefront cannot be generated. The wavefront will be a hemisphere thus reducing the area over which the acoustic power is spread. Since the area of the wavefront is halved, the sound intensity of the wave at that distance is doubled. This is reviewed in the second row of Figure 4.4.

In a manner similar to that described above and as shown in Figure 4.4, rows 3 and 4, a sound source at an edge of two walls radiates into a quarter sphere and has a sound intensity at r of four times the original intensity. Likewise, at a three-wall corner, the source radiates a wavefront that is one-eighth of a sphere and the sound intensity at r for this case is eight times that of the original source in free space.

In some cases, the barrier is not very large with respect to the wavefront; in such cases, the barrier may only partially obstruct sound wave propagation, as shown in Figure 4.5. When this happens, the sound waves will propagate freely through the opening in the barrier. In addition, the barrier's edge at the opening will give rise to the phenomenon of *diffraction.* When diffraction occurs, the edge of a barrier is set in motion by the arriving sound wave and this edge of the barrier acts as a second sound source radiating the sound again (*reradiating* the sound). As with any source, the lower the frequency of the reradiated sound, the more omnidirectional is its radiation pattern. This phenomenon is shown in Figure 4.5. Note from the figure that when a high frequency sound is diffracted, there is a shadow area near the barrier into which no sound is reradiated. Because of this directionality feature of the diffraction process, we find that low frequency sounds show a much greater tendency to "bend around corners" and "pass through small holes in a barrier" than do high frequency sounds.

Having briefly reviewed wave phenomena, we will turn our attention in a later chapter to the analysis of vibrating systems—the type of system we dealt with in this chapter as a simple sound source.

CHAPTER 5
ANALYSIS OF VIBRATING SYSTEMS

One of the primary areas of application for the science of acoustics is the analysis of vibrating systems. After all, making physical and mathematical models is certainly not a useful end in itself. The modeling technique we have considered in earlier chapters are only worthwhile if they help us to understand the operation of real vibrating systems such as musical instruments, sound reproduction equipment and our vocal and hearing mechanisms. As we noted in the last chapter, sound is really the vibration (the periodic movement) of particles of some matter (or sound medium). When a sound wave propagates through the air, the particles of the air are set into vibration such that at different points of time the air particles may be rushing together (compressions) or rushing apart (rarefaction). In order to set the air into vibration, a sound source such as vocal folds or a loudspeaker must vibrate itself. Its vibration will in turn vibrate adjacent particles of air thus beginning the propagation of a sound wave. Sound receivers such as eardrums or microphones must be very sensitive to air particle movement. In this way, the tiny vibrations of air particles will cause the sound receiver to vibrate in the same manner as the air particles, thereby accepting or receiving the sound wave.

Thus, in order to effectively produce or receive sounds, systems must be capable of vibrating. The way in which a system vibrates has a great deal to do with the nature of the sounds it can produce or receive. The acoustician can, therefore, determine how a sound system will operate by analyzing the type of vibrations the system is able to sustain. In this chapter, we will develop a "cookbook" of acoustical modeling steps that can be applied to the analysis of any vibrating system we may wish to study.

ACOUSTICAL PROPERTIES OF VIBRATING SYSTEMS

The acoustical properties of *any* vibrating system are primarily dependent upon the physical attributes of the system that give rise to the system's *inertia* and its *restoring force* components. Since these two properties are so fundamental to a clear understanding of how the system operates, we will consider them individually and in some detail beginning with inertia.

Inertia

Recall from previous chapters that *inertia* is the tendency of a body at rest to remain at rest and of a body in motion to remain in motion. Inertia is a tendency that arises from the *mass* of a body or system. The greater the mass a body has, the greater its inertia—a fact that is well recognized by anyone who has tried to push a broken-down automobile. Clearly, then, it is desirable to know the mass of a given vibrating system. Thus, the acoustician may want to weigh his vibrating system on a scale in order to determine what its acoustical properties might be. Unfortunately, this simple approach to sound system analysis is not entirely adequate. The acoustical properties of a given system depend not only upon the system's mass but upon its size and shape. A dimension that takes both of these features of a system into account is the *density* of the system.

Different types of densities are used in acoustics. The appropriate density for a given body depends upon its shape. Long, thin objects, such as wires and strings, are best described by their length and their density is given as a *linear density*. The linear density of an object is determined by taking the mass of the object and dividing by the length of the object. This value is the mass per unit length of the object or its linear density:

$$\delta = M/\ell \,, \tag{5.1}$$

where

δ = linear density ,
M = mass ,
ℓ = length .

By contrast, large flat surfaces, such as drumhead membranes or flat plates, are best described in terms of their surface area; their density is given as an *area density*. The area density is determined by taking the mass of an object and dividing by the surface area of the object. This value is the mass per unit area of the object or its area density:

$$\alpha = M/a \,, \tag{5.2}$$

where

α = area density ,
M = mass ,
a = area .

Finally, some systems are too thick to be characterized by either linear or area density. Whenever the thickness of the vibrator is large compared to its length and width, the density is calculated as *volume density* (ρ). Vibrators of this sort are usually sound-transmitting media such as air or water. To calculate volume density, we take the mass of the system and divide by its volume (i.e., ρ is mass per unit volume) thus:

$$\rho = M/V \, , \tag{5.3}$$

where

ρ = volume density ,
M = mass ,
V = volume .

Thus, we may determine the inertia tendency of a particular vibrating system by a measurement of one of its physical properties—the density of the system. Depending upon the *shape* of the system, it may be necessary to use either the linear density, the area density or the volume density of the system. Before going into further detail on the use of density to determine the acoustical characteristics of a system, it is necessary to consider the property of systems that gives rise to restoring force.

Restoring Force

Recall from our discussion of pendulum motion that the *restoring force* is that force acting on a system to return it to its equilibrium position. This force acts in opposition to the inertia tendency. By contrast to the pendulum, the restoring force of acoustical systems generally operates independently of the force due to gravity. Acoustical systems, therefore, would operate under zero-gravity conditions while a pendulum would not.

The restoring force of *any* acoustical system arises from the *elastic property* of the system. *Elasticity* is that tendency of an object to return to its original shape. Generally, the greater the elasticity demonstrated by an object, the more rapidly it returns to its original shape. An example of an object with almost no elasticity is a ball made of soft clay. When a clay ball is dropped to the floor, it becomes deformed and stays that way. It splatters to the floor and does not bounce perceptibly. By contrast, a tennis ball is deformed upon hitting the floor but it rapidly returns to its original shape resulting in its bouncing quite well. A tennis ball is moderately elastic. Finally, let us consider a golf ball. It will probably surprise you to learn that a golf ball may also be deformed. High speed photographs of a golf club hitting a golf ball reveal that, upon impact, the golf ball is deformed into a hemisphere for an instant of time. A golf ball, however, is highly elastic and returns so rapidly to its original shape that we are generally unaware that it was ever deformed. It is the high level of elasticity which makes the golf ball bounce so well. Elasticity, therefore, is that property of a body or system which allows it to return to its original shape rapidly.

The elastic property of an acoustical system will arise in one of two ways depends upon the nature of the vibrating part of the system. The vibrator may have a certain amount of stiffness due to its physical makeup, such as a xylophone bar or the plate of a cymbal or air itself. In this case, the *internal stiffness* of the material used provides the restoring force or elasticity of the system. Note that such systems can either be of the type best described by linear density (xylophone bar), by area density (cymbal plate), or by volume density (air).

Some acoustical systems have a vibrator that lacks internal stiffness, such as a guitar string or a drumhead membrane. In these cases, a restoring force will only be present if the vibrator is treated in a specific manner. The vibrator is stretched tightly on some sort of rigid supporting structure. When this is done, the vibrator is under *tension* and it is this tension which gives rise to the restoring force. Note, again, that the system under tension may be one best described by linear density (guitar string) or by area density (drumhead membrane). For purposes of this text, we will not consider the possibility of placing a volume density type of system (such as air or water) under tension.

The restoring force of an acoustical system, therefore, arises from its elasticity. This elasticity may arise either from the internal stiffness or the tension present in the system depending upon the system's physical properties. As we have seen in Chapters 2 and 3 (using the pendulum model), the interactions of inertia and restoring force exert considerable control over system operation. Thus, by determining the density and elasticity of an acoustical system, we can calculate several of its basic properties.

Friction

The reader will recall from Chapters 2 and 3 that there is one other force (in addition to the restoring force) that exerts some control over system operation. This other force is *friction*. When we study friction in a vibrating system, we are usually studying the friction generated by particles of the system (be it a string, a plate, air or any other system) as they rub against one another when the system vibrates. So, the frictional force of interest for acoustical systems arises internally within the vibrator. The material property associated with this force is the *damping* of the system. Damping of acoustical systems operates in the same manner as that described in Chapter 4 of this text. Recall, then, that the time required for a system to reduce its velocity amplitude to 0.37 of its original amount depends upon the magnitude of the damping. The greater the amount of damping, the more rapidly the motion of the system dies away. Thus, one measurable property of the system that would indicate the magnitude of the damping present is a measure of the time required for some motional component of the system to decrease in amplitude by 63%. This measurement is referred to as the *decay time* of the system and it is a method used to characterize the damping of any type of acoustical system.

Summary: Acoustical Properties of Vibrating Systems

The mass, elasticity and damping properties of a vibrating system are associated with specific forms of motion and energy demonstrated by the system. The *total force* arises from the *mass* of the system and is dependent upon the system's *acceleration* (F = ma—Newton's second law, Chapter 2). Recall also that the *inertial force* gives the system its tendency to maintain a uniform state of motion (i.e., resting body wants to stay at rest and moving body wants to keep moving). Thus, the total force is associated with the energy of motion—*kinetic energy*.

In opposition to inertia is *restoring force*—that force which makes the particles of the system move towards their equilibrium positions. In effect, this force limits the extent to which a particle may be removed from its equilibrium position. This is to say that the restoring force controls the *displacement* component of the system's motion. Note that restoring force really acts upon a particle of the system when it is displaced from equilibrium. In this respect, restoring force is closely related to the position of the particle of the system upon which it is acting. Thus, we may see that restoring force is associated with the energy of position—*potential energy*.

Finally, *frictional force* acts upon the vibrating system in the form of *damping*. As noted in Chapter 2, friction is associated with *dissipative energy* through heat generation. When the particles of a system are set into vibration, they rub against one another, generating heat. This rubbing process presents a definite resistance to particle motion within the system and thereby limits the rate at which particles move. Frictional force thus limits the *velocity* component of motion and is associated with the energy of heat—dissipative energy.

As a summary for those matters we have discussed in this chapter so far, Table 5.1 is provided. It demonstrates the forces operating in an acoustical system along with related features of the system itself and its forms of motion and energy. Having completed this discussion, we will now proceed to describe the specific steps to follow in the analysis of any vibrating system.

STEPS IN ACOUSTICAL ANALYSIS

1. *Identify the material properties of the system and determine how they may best be measured or found.*

All systems have three material properties with which we are concerned in the field of acoustics—mass, elasticity and damping. The type of system with which we are dealing may be identified by noting the specific manner in which these material properties arise or are best expressed. In all systems, the internal damping is best defined in terms of decay time. Thus, damping is essentially the same for all types of systems. Note that it is an especially useful property for determining how long a system's vibration will persist (see Chapter 2) and how sharp a system's resonance peak is (see Chapter 3). Aside from this,

Table 5.1.
Fundamental Relationships Observed in Vibrating Systems.

Force	*Material Property*	*Measurable Feature*	*Motional Component*	*Energy*
Total	Mass	Density	Acceleration	Kinetic
Restoring	Elasticity	Tension or Stiffness	Displacement	Potential
Frictional	Damping	Decay Modulus	Velocity	Dissipative

we will no longer be concerned with damping in this discussion of system analysis.

The *elasticity* present in the system may arise from either of two measurable features of the system—stiffness or tension. Thus, all of the systems in which we have an interest may be placed in one or two categories. In this chapter, we will consider six model vibrating systems that represent all of the basic types of acoustical systems you may wish to analyze. By simply selecting the system that is most similar to the system we wish to study, we can identify the formulas that control this system and, thereby, calculate its properties. The systems we will use as models are the air column, the resonator, the string, the membrane, the bar, and the plate. Those systems having an elasticity arising from internal stiffness include the air column, the resonator, the bar, and the plate. The string and the membrane, however, will only vibrate when they are stretched, so they derive their elasticity from tension. Thus, we can begin identifying the most reasonable acoustic model for any system we wish to study by noting the manner in which the elasticity arises.

A second means of identifying the type of vibrating system we are studying is to identify the most reasonable means of describing the *density* of the system. The student will recall that density may be characterized in either linear, area, or volumetric terms. Considering the six basic models we have named, we find that the string and the bar are long and thin and are best described in terms of a linear density. On the other hand, the membrane and the plate are large flat surfaces and are best described in terms of an area density. Finally, the air column (such as an organ pipe) and the resonator (such as a bottle you blow into) have both width and thickness as well as length and are best characterized by their volume density.

The model which best describes a specific acoustic system then may be identified by simply considering the source of the elasticity and the nature of the density of the system being studied. As an aid in acoustic system modeling, Table 5.2 is provided. To use the table, select the appropriate row by identifying the nature of the system's density. Then, noting the type of elasticity for the system, select the appropriate column. This leads to a specific box in the table and identifies the appropriate acoustic model for the system of interest.

2. *Determine the type of wave motion sustained by the system and what type of wave we wish to study.*

It is interesting to observe that the top box in the stiffness column has two entries—the bar and the beam. This is because the linear

Table 5.2.
Selection of Acoustic System Models.

		Elasticity	
		Stiffness	Tension
Density	Linear δ	Bar or Beam	String
	Area σ	Plate	Membrane
	Volume ρ	Air Column or Resonator	--------------------

density type of system that has internal stiffness may sustain any of three types of sound waves—transverse, longitudinal or torsional. In fact, when such a system is excited, all three types of waves are excited although one type typically predominates depending upon how the bar is excited. When a force is applied at the end of the system, the longitudinal mode of vibration predominates. When the force is applied across the bar along a side, the major vibration is transverse; and when the force is a twisting one, the predominant mode of vibration is torsional. For further discussion of wave types, see Chapter 4. When we study this type of system and concentrate on its longitudinal vibration pattern, we refer to the system as a bar, whereas the system is called a beam when the vibrational mode of interest is either transverse or torsional. The acoustical analysis of the system varies depending upon whether we are interested in its bar-like or its beam-like properties. In this text, we will consider bar vibration. Beam vibration is rather more complex and is not within the scope of this text.

3. Identify the boundary conditions of the system.

Note that the bottom box in the stiffness column of Table 5.2 has two entries—air columns and resonators. This is due to a special feature of systems that derive their elasticity from internal stiffness. A system that is under tension (such as a string) must be tightly and immovably clamped at its edges or ends in order for that tension to exist. A loosely attached guitar string makes no sound (does not vibrate). It only vibrates when it is attached tightly to the guitar at its ends. This means that the ends of a guitar string cannot move. Any limitation on system movement due to the method of supporting the system is referred to as a *boundary condition*. The boundary conditions of all systems that derive their elasticity from tension require that the system not vibrate at its ends or edges. Any point in a system that is in some way prevented from vibrating is referred to as a *node*. Thus, we may say that the boundary conditions of a string are such that it must have nodes at its ends. A schematic drawing of string vibration at its natural frequency is given in Figure 5.1. Note the nodes at the points of attachment of the string. The dashed lines represent the upward and downward limits of the string's movement as its vibrates. Note that the part of the string that moves with the greatest displacement amplitude is at the center of the string, halfway between the two nodes. The point on the vibrating system that shows the greatest displacement amplitude is called an *antinode*.

By contrast with the systems having elasticity due to tension, the systems having internal stiffness need not have any fixed boundary conditions. Since the restoring force is due to stiffness, it will be present whether the edges of the system are tightly attached or not. This system need only be clamped so as to hold it in place; the clamp may be applied at any point on the system. So, we may say that the boundary conditions are invariant for tension-type systems while the boundary conditions are variable for stiffness-type systems. Let us consider the possibilities for an air column type of system. Recall that the vibration of air is

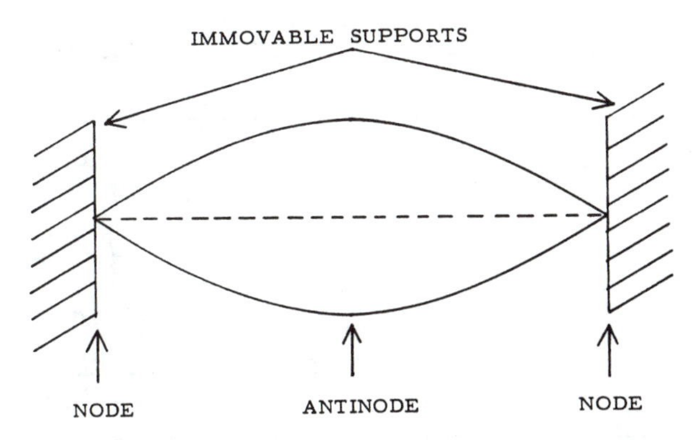

Figure 5.1. Fundamental vibration pattern of a string. Vibration is transverse.

longitudinal rather than transverse as with the string. Whenever we try to characterize the longitudinal motion of a vibrator, however, it is easiest to draw a graph of the displacement amplitude superimposed on the sketch of the system. Unfortunately, this type of graphic looks like a transverse vibration, so difficulty in interpretation sometimes arises. If the reader will try to keep in mind that the graph represents the displacement amplitude of a longitudinal motion by its distance from the axis of the graph, there should be no problem. The air column is a pipe filled with air such as an organ pipe. The air in the pipe can be set into vibration in the same manner as any other vibrating system. Figure 5.2 shows the vibration patterns of an air column and demonstrates the

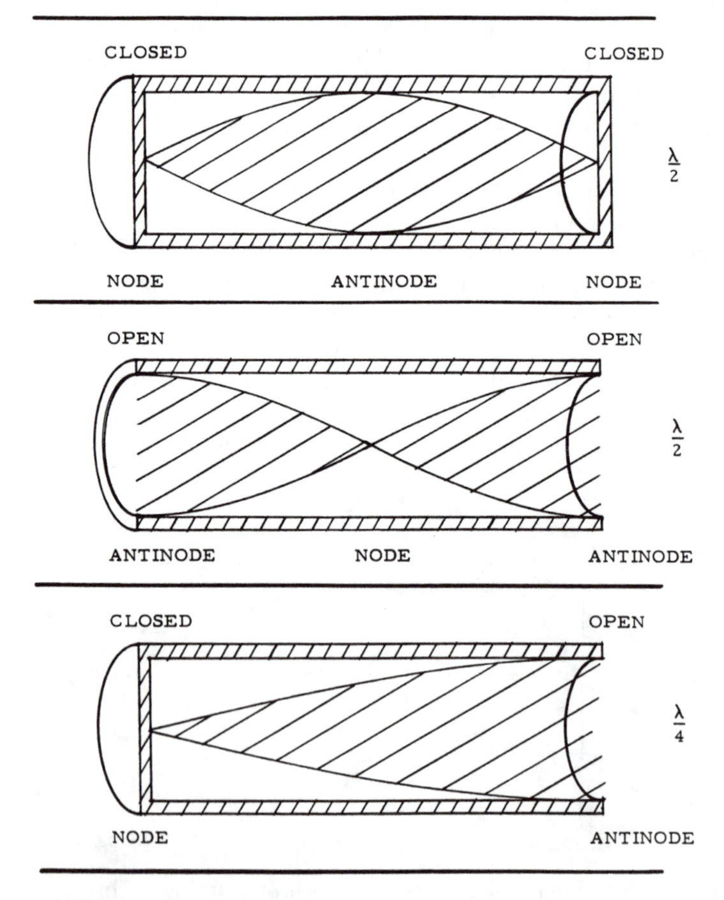

Figure 5.2. Pressure distribution in an air column.

effect of varying the boundary conditions on the air column. When the end of the air column is closed, there is a rigid wall at the end of the pipe and no vibration can occur there. Thus, the boundary condition of a closed end necessitates the presence of a node at the end. By contrast, when the end is open the air is allowed to vibrate freely, so the open boundary condition requires that an antinode be present at the end of the pipe. A brief examination of Figure 5.2 shows that varying the boundary conditions of a given system will result in quite different types of sound wave propagation. In the first case, there are nodes at both ends of the system and an antinode at the center as with the string. In the case with both ends open, there are antinodes at the ends of the pipe and a node at the center. Note, however, that in both these cases the distance between each node and antinode is the same—half the length of the pipe. Integrating our discussion of periodic motion in Chapter 2 with Figure 5.2 reveals that one cycle (having a length of one wavelength—λ) is four times the distance between any node and an adjacent antinode. So, we may state that the wavelength of the sound at the natural frequency of the pipe is:

$$\lambda = (4) \times (1/2\,\ell)\,,$$

or

$$\lambda = 2\ell\,, \tag{5.4}$$

where

λ = wavelength, and
ℓ = the length of the pipe .

By contrast, in the open-closed pipe there is a node at the closed end and an antinode at the open end. So the distance between the node and the adjacent antinode is equal to the length of the pipe and we may say that, for the closed-open pipe:

$$\lambda = (4) \times (\ell)\,,$$

or

$$\lambda = 4\ell\,. \tag{5.5}$$

Clearly, then, the boundary conditions can change the wavelength of the natural frequency of the system. It is, therefore, necessary to define the boundary conditions of the system in order to analyze its operation.

The boundary conditions for the various systems in which we have an interest as well as the wavelength of the natural frequency of the system are given in Table 5.3. There are several notations used in this table which will require explanation.

The second model entered in the table is the *Helmholtz resonator* or simply the resonator. The *resonator* is simply an enclosed volume of air having a port at some point. A common example of such a system is a bottle. Blowing across the opening of a bottle or jug produces a sound at its resonance frequency. What actually happens is that a small mass of air in the opening of the bottle moves up and down through that opening. It is pushed down by the driving force (i.e., your breath as you blow) and it is forced back upwards by the elasticity of the air enclosed in the bottle. Figure 5.3 shows a diagram of a resonator identifying the material aspects of the resonator used to compute the wavelength of its vibration.

The other notation used in Figure 5.3 is used to determine the wavelength of the sound wave in a membrane. The necessary measurement for this calculation is the radius of the membrane (a).

4. Determine the speed of sound in the system of interest.

One of the primary timing aspects of a sound transporting system is the speed of sound in the system. In order to analyze the operation of vibrating systems, we must consider the speed at which sound travels in these systems. The fact that sound travels faster in some materials than in others has been recognized even by primitive peoples. We have all seen western movies where Indians are shown listening for the approach of the calvary by placing their ear to the ground. The approaching horses can be heard through the ground before they can be heard through the air. This is because sound travels faster through the Earth than through the air. The speed of sound (c) in any matter is dependent upon two material properties of that matter—its elasticity and its density. The specific mathematical relationship is shown in the general equation below:

$$c = \sqrt{\frac{\text{elasticity}}{\text{density}}}\,. \tag{5.6}$$

As we have discussed earlier, the elasticity and density features of the system under study should be identified at the outset of the analysis. The density of the system may be given as a linear, area or volume density as designated in Table 5.3. The elasticity may be a measurable tension or a stiffness (again noted in Table 5.3). The stiffness of solids, such as bars, beams and plates, is given by the *elasticity modulus* or

Table 5.3.
Summary of Vibrating Systems.
System Analysis Summary Table

Model	*Material Property* *Elasticity*	*Density*	$c =$	*Boundary Conditions*	$\lambda =$	$f_n =$
Air Column	γp_0	ρ_0	$\sqrt{\frac{\gamma p_0}{\rho_0}}$	open-open closed-closed	2ℓ	$\frac{nc}{2\ell}$
Air Column	γp_0	ρ_0	$\sqrt{\frac{\lambda p_0}{\rho_0}}$	closed-open	4ℓ	$\frac{(2n-1)c}{4\ell}$
Resonator	γp_0	ρ_0	$\sqrt{\frac{\gamma p_0}{\rho_0}}$	with port	$\frac{2\pi}{\sqrt{S/\ell V}}$	$\frac{c}{2\pi}\sqrt{\frac{S}{\ell V}}$ *
String	T	δ	$\sqrt{\frac{T}{\delta}}$	fixed ends	2ℓ	$\frac{nc}{2\ell}$
Membrane	T	σ	$\sqrt{\frac{T}{\sigma}}$	fixed edges	$\frac{2\pi a}{2.405}$	$\frac{2.405}{2\pi a}\sqrt{\frac{T}{\sigma}}$ *
Bar	Y	ρ	$\sqrt{Y/\rho}$	free-free	2ℓ	$\frac{nc}{2\ell}$
Bar	Y	ρ	$\sqrt{Y/\rho}$	fixed-free	4ℓ	$\frac{(2n-1)c}{4\ell}$

Bar with torsional or transverse waves not given—Complex

Plate not given—Complex

*Inharmonic Series

Note: This appears in Appendix A, Study Guide, as Table A.5.1.

$$\lambda = \frac{2\pi}{\sqrt{S/\ell V}}$$

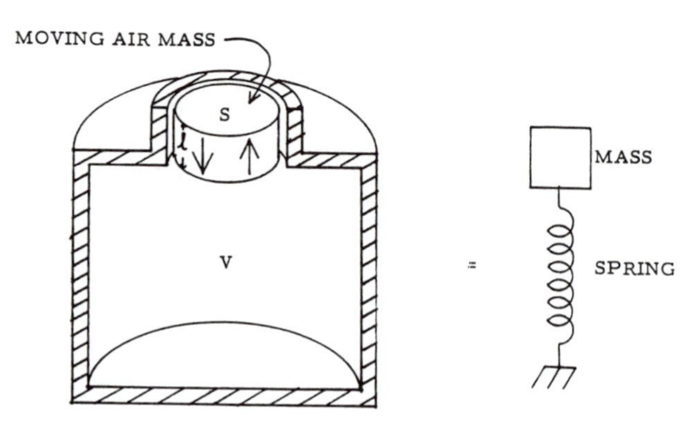

V = Volume of enclosed air

S = Surface area of neck opening

ℓ = Length of neck opening

Figure 5.3. Schematic diagram of a Helmholtz resonator and equivalent mass—spring system.

Young's modulus and there are published tables of the Young's modulus of various solids. The stiffness of gases, such as air, is given by the product of two quantities:

$$K = \nu P_0 \,, \tag{5.7}$$

where

K = the stiffness of the gas,
ν = the elasticity of the gas which varies with temperature, and
P_0 = the ambient (or background) pressure of the gas which varies with altitude.

Thus, by substituting the appropriate terms into the general equation given above, it is possible to determine the speed of sound (c) in any system of interest.

5. Determine the resonance or natural frequency of the system.

In the last two steps we have determined the distance between one wavefront and the next, the wavelength (λ), and we have calculated the speed of sound (c) in the system of interest. Recall for a moment from your high school mathematics the formula:

$$\text{distance} = \text{rate} \times \text{time} \,,$$

or

$$d = rt \,. \tag{5.8}$$

We have already noted that frequency is inversely proportional to period (a time measure):

$$T = 1/f \,, \tag{5.9}$$

so we could rewrite our formula:

$$d = r/f \,. \tag{5.10}$$

The distance we are interested in is the wavelength (λ) and the rate we are interested in is the speed of sound (c). We can, therefore, rewrite the equation again as:

$$\lambda = c/f \,. \tag{5.11}$$

Rearranging our equation so as to solve for frequency, we see that:

$$f = c/\lambda \,. \tag{5.12}$$

By substituting the values from the previous two steps of our analysis into this equation, we may determine the resonance frequency of any system of interest.

6. Determine the overtones of the system.

The *overtones* of a system are the other natural frequencies of the system. The overtones are higher in frequency than the resonance frequency and usually have lower amplitude peaks than the resonance frequency. In most systems, they are excited whenever the resonance frequency is generated. There are three types of overtone series that may be observed in acoustic systems. Symmetrical, linear-density type systems sustain harmonic series. These systems include: closed-closed air columns, open-open air columns, strings and free-free bars. A *harmonic series* is one in which the overtones are integral (whole number) multiples of the resonance frequency. The formula for computing harmonic overtones is:

$$f_n = n \times f_1 , \tag{5.13}$$

where

f_1 = the resonance frequency,and
n = the number of the overtone.

Unsymmetrical, linear-density type systems sustain odd-harmonic series. These systems include the closed-open air column and the fixed-free bar. An *odd-harmonic series* is one in which the overtones are odd, whole-number multiples of the resonance frequency. The formula for computing odd-harmonic overtones is:

$$f_n = (2n - 1)f_1 . \tag{5.14}$$

Finally, the remaining systems including the resonator, the membrane, the beam and the plate sustain inharmonic series. An *inharmonic series* is one in which the overtones are not integral multiples of the resonance frequency. It is beyond the scope of these notes to explain how to calculate inharmonic series. The reader is advised to seek out the tabled values of f_n for the various inharmonic sources. These data are available in source books in acoustics.

CHAPTER 6
SPEECH PRODUCTION

THE LARYNX

The primary function of the *larynx* is to protect the lungs from food particles that could cause suffocation. The *epiglottis* (see Figure 6.1) can be angled to guide food into the *esophagus*, preventing food from entering the larynx. Should some food particles get past the epiglottis, the vocal folds will close to prevent food from getting into the *trachea*. The coughing reflex can then blow this food out of the larynx so it can be expelled from the body. By careful manipulation of the protective mechanisms of the larynx, sounds can be produced. These sounds form the basis of what we call speech. The process of producing sound at the larynx for speech is called *phonation*.

The *glottis* (the opening between the vocal folds) can be closed during exhalation. You do this by pulling your *vocal folds* strongly together as you do when food enters your larynx. If your glottis is closed when you are trying to exhale, the air flow will stop at your larynx and will not get out of your body. If you keep pushing with your muscles of exhalation, the air pressure within the *sub-glottal space* (superior part of the trachea just beneath the glottis) will increase. Eventually, the force of the air pressure in the sub-glottal space will be greater than the force of the muscles holding your glottis closed. Then the air pressure will blow the vocal folds apart and a puff of air will be released from the sub-glottal space. This air puff will travel through the *laryngopharynx* and *oropharynx* to the oral cavity to be released from the body at the mouth.

When air pressure from the sub-glottal space blows the vocal folds apart, a lot begins to happen. First, the release of the puff of air briefly reduces the amount of air in the sub-glottal space. This reduces the air pressure in the sub-glottal space so that it is no longer enough to blow the vocal folds apart. Because of this reduced air pressure, the glottis begins to close immediately after opening. A second effect, called the Bernoulli effect, pushes the vocal folds together with even more force.

The *Bernoulli effect* states that whenever a mass moves quickly, there will be a space right behind it where the air pressure is very low.

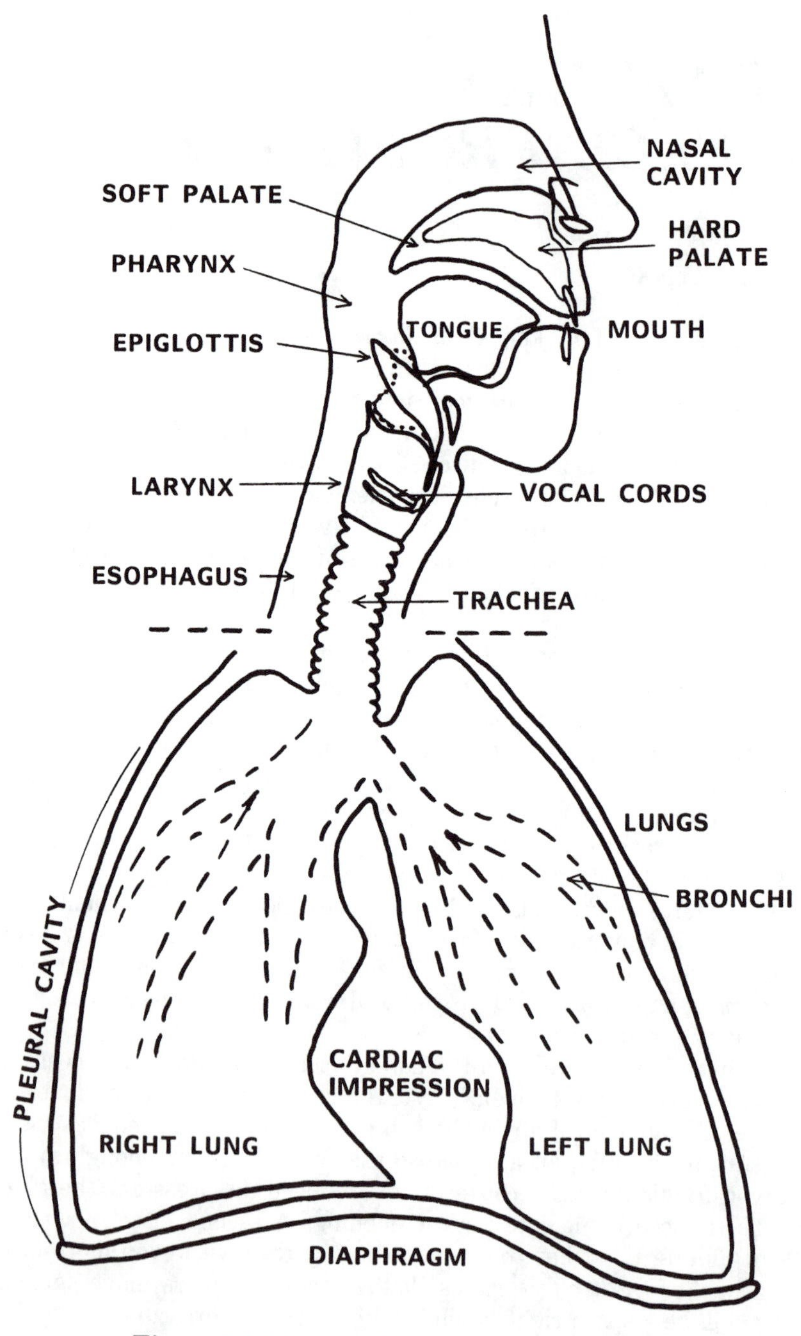

Figure 6.1 The speech production mechanism.

There is a reason for this Bernoulli effect. When a mass moves through the air, it pushes air out of its way, both ahead of itself and to the sides. Once the mass has gone by, the air that has been pushed aside returns to fill in the space behind the moving mass. This takes some time, however, and there is a place right behind the moving mass that has fewer air particles than the rest of the air around the mass. That is, there is a space of low air pressure just behind the moving mass. As air moves to fill in this low pressure space, solid objects may be blown towards the space behind the moving mass. The Bernoulli effect can be seen in action on autumn days when leaves get blown towards the space behind a moving car when it goes by.

The Bernoulli effect acts on the vocal folds. When a mass of air is released from the sub-glottal space, there is a low pressure area just behind the moving air puff. This low pressure area passes through the glottis just after the air puff and causes the vocal folds to be pulled together to fill the low pressure area. So, the Bernoulli effect along with the reduced air pressure in the sub-glottal space causes the vocal folds to be drawn together, closing the glottis after the release of an air puff.

Once the glottis has closed again, air pressure begins to build up once more in the sub-glottal space, and the whole process is repeated. This process of releasing air puffs and closing the glottis will continue so long as the exhalation muscles can continue to cause an air pressure buildup in the sub-glottal space. When the larynx is used in this way, it can release a series of air puffs. In acoustics, a device which produces sound by releasing a series of air puffs is called a *siren*.

Before continuing, let us make a simple siren so we can understand the principle of siren operation. Get a small balloon and inflate it. Now, hold the neck of the balloon stretched between thumb and index finger of each of your hands. As the air escapes from your balloon, it does so by blowing apart the folds of the balloon's neck. After each puff of air escapes, the folds close briefly. Then, the folds are blown apart by the pressure of the deflating balloon and another puff escapes. The balloon's neck acts as a larynx in our model. The balloon itself acts as the muscles of exhalation. Your thumbs and fingers act as the muscles which pull the vocal folds together, and the folds of the balloon's neck act as vocal folds.

Notice that as the siren's air puffs escape they create a sound. This happens because the fast moving puffs of air released by the siren run into the still air near the siren's neck. The sound you hear is the sound of air running into air. This principle is the same one that leads to phonation at the human larynx. The frequency of the sound created by a siren depends on the rate at which it releases air puffs. If 135 air puffs are released each second, the siren produces a 135 Hz sound. As with our balloon model, the frequency of the phonation produced by your larynx depends on the number of air puffs it produces each second.

THE VOCAL FOLDS

Before examining the anatomy of the larynx, we will look at the acoustics of vocal fold operation. The vocal folds are muscular tissues that can be stretched across the glottis, closing the opening from the pharynx to the trachea. The vocal folds are made up of two components: a muscular band stretched from anterior to posterior within the larynx, and a member which extends from the muscular band to the lateral wall of the larynx where it is sealed (see Figure 6.2). When the muscular bands are pulled together, the glottis is closed. When the vocal folds close the glottis, they can be blown apart to produce puffs of air as we saw in the last section. The number of puffs released per second controls the frequency of the phonation produced and the acoustical properties of the vocal folds control their rate of releasing air puffs.

We can consider vocal fold operation by modeling the vocal folds as a simple acoustical system. Slow motion films of vocal fold movement show that is is the muscular bands of the vocal folds that move during phonation. In fact, the vocal folds do not move apart as a single unit, but blow apart at a posterior point along their muscular edges. This opening of the glottis then moves slowly along the vocal fold edge until it reaches the anterior end of the vocal fold. Then the opening moves back toward the posterior end of the vocal fold where the process begins again. This characteristic movement along a vibrating system is a wave. For a review of wave phenomena see Chapter 4. The vocal folds, then, are an acoustical system made up of bands of tissue stretched across the inside of the larynx. They can be acoustically modeled as a vibrating string.

Let us see how our string model of the vocal folds works by calculating the resonance frequency of the model and comparing it to the known phonation frequency of the human vocal folds. A human male adult vocal fold is about 1.8 cm (0.018 m) long and has a linear density of about 0.127 kilograms per meter. If the vocal folds are stretched to a tension of about 3 newtons the speed of sound in the vocal folds would be:

$$c = \sqrt{T/\delta} = \sqrt{3.0/.127} = \sqrt{23.622} = 4.86 \text{ m/sec}$$

So the speed of sound in the adult male vocal fold is about 4.86 m per sec. Substituting this value and the wavelength of 0.036 m (2 × 0.018 = 0.036), we have:

$$f = (4.86 \text{ m per sec})/(0.036 \text{ m}) = 135 \text{ per sec}$$

and the model vocal fold has a resonance frequency of 135 hertz. This compares well with measured vocal fold operating frequencies for males.

In Chapters 4 and 5 we noted that one unique aspect of continuous systems is that they have more than one resonance frequency. The frequency of the system's simplest vibration mode is called the funda-

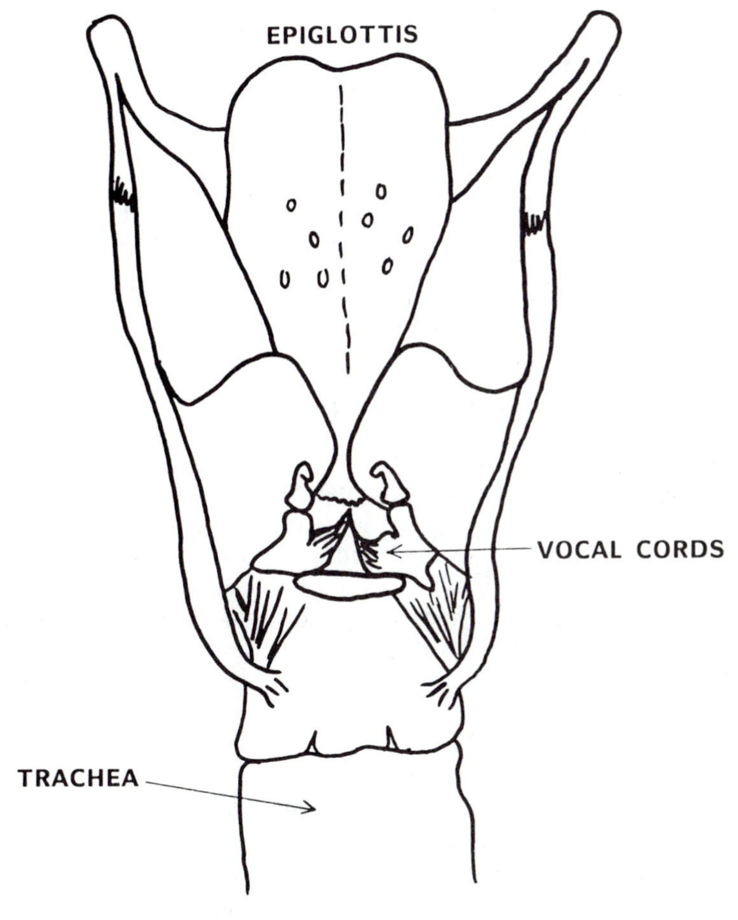

Figure 6.2 Rear view of the larynx.

mental resonance frequency of the system, while the frequencies of the other resonances are called overtones of the fundamental frequency of the system. Let us look at the pattern of these overtones for the vocal fold model we just calculated. The second resonance (first overtone) frequency would be:

$$f_2 = 2(135 \text{ Hz}) = 270 \text{ Hz}$$

The third resonance (second overtone) frequency would be:

$$f_3 = 3(135 \text{ Hz}) = 405 \text{ Hz}$$

The fourth resonance (third overtone) frequency would be:

$$f_4 = 4(135 \text{ Hz}) = 540 \text{ Hz}$$

Likewise, other overtones would include 810 Hz, 1080 Hz, 1350 Hz, 1620 Hz and so forth.

What if the vocal fold's tension were raised to 12 N (four times the original tension)? This would change the speed of sound:

$$c = \sqrt{12/.127} = \sqrt{94.49} = 9.72 \text{ m/sec}$$

This doubles the original fundamental resonance of the vocal folds. As we noted in Chapter 5, the speed of sound increases as the square root of tension. The new fundamental resonance frequency will be:

$$f = (9.72 \text{ m per sec})/(0.036 \text{ m}) = 270 \text{ Hz}$$

Note that the new fundamental resonance frequency is double the original fundamental resonance frequency. Since frequency is directly dependent on speed of sound in a material, doubling the speed of sound doubled the frequency, as we saw with the string model in Chapter 5. Doubling the frequency is an increase in voice pitch of one octave. A good singer can vary his or her voice pitch over a range of three octaves. That is sixteen times the vocal fold tension (from 3 N to 48 N). Some professional singers can vary their voice pitches almost four octaves. That is an increase in vocal fold tension of 25 (from 3N to 75 N)!

An adult female vocal fold has about the same linear density and average tension as a male vocal fold. It is only about 1.1 cm (0.011 m) long, however. Try calculating the fundamental resonance frequency of a female adult vocal fold. It should be about 220 Hz. Also try calculating the overtone frequencies of a female vocal fold. They should be about 440 Hz, 660 Hz, 880 Hz, 1100 Hz, 1320 Hz, and so forth.

MODES OF PHONATION

A generally accepted theory of speech production is the *source-filter theory* in which the character of a complex noise source (for example the siren discussed earlier in this chapter) is modified by passing through a series of filters. Let us examine the process of phonation with this theory in mind. So far we have talked about the larynx as though it were either phonating or not. In fact, however, there are several *modes of phonation* and in this section, we will look at these various different operating patterns for producing speech sounds.

The most common pattern of phonation is called normal voicing. In *normal voicing* (or *voiced speech*) the vocal folds are adducted with a moderate amount of vocal fold tension. By varying the vocal fold tension, a speaker can adjust voice pitch as we discussed in the section on vocal fold acoustics. The siren-like behavior of the larynx during normal voicing produces a hum or buzz at the larynx. This sound can be adjusted by the other parts of the vocal tract to produce speech sounds.

Some speech sounds are produced without closing the glottis. These are called *voiceless speech*. In voiceless speech, the vocal folds are completely abducted and air passes freely and soundlessly through the glottis. Voiceless speech sounds are produced by creating some configuration of the vocal tract that causes noise at a point superior to the larynx. For example, the sound /s/ (as in "sing") is produced by constricting the airflow through the mouth with the tongue causing a friction noise. The same mouth and tongue configuration can be used with normal voicing to produce /z/ (as in "zoo"). Two speech sounds that are produced in the same way and differ only in whether or not normal voicing is used are called *cognates*. So, /z/ is the *voiced cognate* (cognate with normal voicing) of /s/. By contrast, /s/ is the *voiceless cognate* (cognate without normal voicing) of /z/. Other cognate pairs include /f/ which is voiceless and /v/ which is voiced; as well as the voiceless /p/ and the voiced /b/.

In addition to these two extremes of voiced and voiceless speech, there are numerous intermediate modes of vocal fold operation. *whispered speech* is produced by narrowly opening the vocal folds. Air is allowed to flow through the narrow opening but, because the glottis is narrowed, the airflow is noisy. In acoustics, this phenomenon is called *jet noise* and is acoustically similar to the sound produced by jet aircraft. When an airflow is forced through a narrow opening, the air moves very quickly through the opening. This is the only way to get the same amount of air flowing as we would have if the opening were wide. When this high speed air comes out of the glottis, it rubs against the still and slow moving air above the glottis and we hear a friction noise caused by the rubbing of air particles moving at radically different velocities. This phenomenon results in the characteristic hissing sound we call whispered speech.

Breathy speech occurs when the vocal folds are adducted anteriorly and narrowly abducted posteriorly. The airflow rushes quickly through the narrow, posterior opening of the folds causing a jet noise, as in whispered speech. The anterior vocal folds release air puffs in a siren-like manner, as in normal voicing. This combination of voicing with friction noise is called breathy speech.

A *creak* (or *vocal fry*) type of sound is produced when thick, relaxed vocal folds are pulled tightly together. When this occurs, air puffs are released irregularly and slowly from the center of the glottis. The resulting sound is the one most people make when they try to imitate a "creaky old door." A vocal fry is generally characterized by the popping sound of the vocal folds striking one another during production of this type of sound. Generally, the individual pulses of the vocal fry sound are heard individually because they are far enough apart in time for us to hear silent intervals between the pulses. The pulses of siren-like sound produced during normal voicing are perceived as a hum because they are too close together in time to be heard individually.

VOCAL RESONANCE

Remember we said that any system that vibrates will have a natural vibrating frequency? This natural frequency is called the resonance frequency of the system. Most systems are made up of many interconnected components. These are continuous systems and may have many resonance frequencies. We have noted that this is the case for vocal fold vibration. For continuous systems, the lowest resonance frequency is called the *fundamental frequency* of the system. Other resonance frequencies of a system are called overtones of the fundamental. By contrast, some systems operate as a single unit and are called discrete systems. This was the case for the bottle in Chapter 5. Discrete systems have only one resonance frequency. For additional information, see Chapters 4 and 5.

There are three important internal forces in any acoustical system. These are the stiffness, the momentum, and the friction of the system. Together these forces combine to control the way the system operates. They determine the system's fundamental and overtone resonance frequencies as well as its response amplitudes, both at and away from resonance, for any applied driving force. Characteristic system impedance is a useful way of describing how these three forces will interact in a given acoustical system. Impedance describes the reluctance of a system to move in response to a particular driving force. By determining the magnitude and phasing of the three forces for a particular system, we can mathematically predict how the system will operate. By modeling real systems in simple ways (as we did for the vocal folds), we can predict how they will operate. (For a review of these types of predictions, see the discussion of Q in Chapter 3 and the samples of resonance calculation in Chapter 5.)

THE VOCAL TRACT

The *vocal tract* is a tubular pathway that extends from the larynx through the neck and the oral cavity to the mouth. Muscular adjustment of this tract can alter the phonation from the larynx to produce specific speech sounds. The vocal tract includes the laryngopharnyx, the oropharynx, the oral cavity, and the mouth. These four structures are, in fact, respiratory structures, but they also serve as resonating structures in the speech production process. Be careful about what you call these four! Some working in speech science call these "resonators". While these structures do resonate, calling them resonators is technically incorrect. A resonator is a discrete system, like a bottle or automobile muffler. The components of the vocal tract interact as a continuous system and, as such, they sustain numerous resonances simultaneously. The vocal tract is best modeled as an air column, a continuous system. It is not an acoustical resonator!

Let us examine the acoustical operation of the vocal tract during speech production. In the following discussion, please refer to the schematic diagram of the vocal tract, Figure 6.3. The most important single factor in air column operation is the length of the air column. This controls the wavelength of the resonance frequencies of the air column. In general, the vocal tract is about 24 cm long. The second most important factor in air column vibration is boundary condition. The boundary conditions determine whether a node or an antinode will occur at the end of a system. The boundary conditions of the vocal tract can be varied by the speaker, so we will have to look at more than one boundary condition pattern for the vocal tract in our analysis. (See the discussion of air column analysis in Chapter 5.)

When a person is phonating, the vocal folds are blown apart

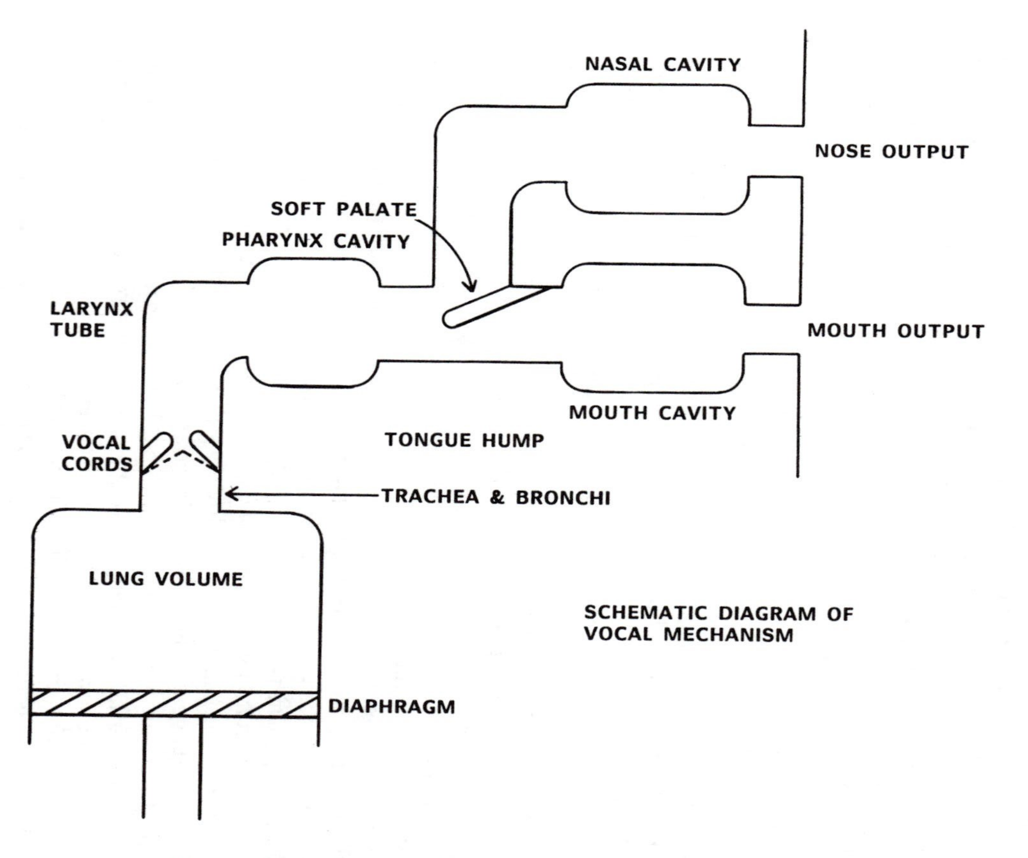

Figure 6.3 Schematic diagram of the vocal mechanism.

periodically during the release of air puffs. Though they are generally in a closed position between air puffs, they are apart when sound enters the vocal tract. Acoustically, therefore, we say that the vocal folds form a free boundary condition at the laryngeal end of the vocal tract. By contrast, the boundary condition of the mouth end of the vocal tract can be greatly varied during phonation. For a vowel like /a/ (as in the word "father," the mouth is wide open and the boundary is free. For a vowel like /u/ (as in the word "glue"), however, the mouth is almost closed, greatly restricting air particle movement and approximating a fixed boundary. We will look at these one at a time to see how the vocal tract operates in each case.

First let us look at the vowel /u/ (as in "glue") where the mouth is almost closed and approximates a fixed boundary. Obviously, the boundary cannot be completely fixed (mouth completely closed) because no sound could get out. The lip position for /u/ is, however, very close to being closed and can be modeled by treating it as if it were closed. We have already said that the laryngeal end of the vocal tract is free during phonation. So, during the production of the vowel /u/, the vocal tract has a free laryngeal boundary and a fixed mouth boundary. In Chapter 5 we said that for unsymmetrical systems the wavelength of the fundamental frequency is four times the length of the system. Because the vocal tract is about 24 cm long, the wavelength of the first mode of /u/ should be:

$$\lambda = 4 \times 24 \text{ cm} = 96 \text{ cm}$$

So the wavelength of the unsymmetrical vocal tract phonating the vowel /u/ is about 96 cm or about 0.96 m. From this basic value and the speed of sound in air (343 m per sec, from Chapter 5), we can calculate the fundamental resonance frequency of the vocal tract using the wavelength formula from Chapter 5. Recall the formula

$$f = c/\lambda,$$

where:

λ = wavelength
c = speed of sound, and
f = frequency

This would result in a fundamental resonance frequency of:

$$f = (343 \text{ m per sec}) \,/\, (0.96 \text{ m}) = 357 \text{ per sec}$$

The unsymmetrical vocal tract phonating the vowel /u/ should have a resonance frequency of about 357 hertz. Measurements of adult phonation of the vowel /u/ result in a fundamental frequency of about 335 Hz, so our model is close.

Now let us look at the overtones of our unsymmetrical vocal tract producing the vowel /u/. Recall from Chapter 5 that the overtones of an unsymmetrical system are odd number multiples of the fundamental, that is:

$$f_2 = 3\,(f_1)$$

and,

$$f_3 = 5(f_1)$$

or:

$$f_n = (2n - 1)\,(f_1)$$

So the second resonance is:

$$\begin{aligned} f_2 &= [(2 \times 2) - 1]\,(357 \text{ Hz}) \\ &= (4 - 1)\,(357 \text{ Hz}) = 3 \times 357 \text{ Hz} = 1071 \text{ Hz} \end{aligned}$$

Laboratory measurements of the first overtone of the vowel /u/ show values around 1000 Hz. So we are still close. The third resonance of the vocal tract phonating /u/ should be:

$$\begin{aligned} f_3 &= [(2 \times 3) - 1]\,(357 \text{ Hz}) \\ &= (6 - 1)\,(357 \text{ Hz}) = 5\,(357 \text{ Hz}) = 1785 \text{ Hz} \end{aligned}$$

Laboratory measurements of the second overtone of /u/ are near 2000 Hz, a bit higher than what we calculated, but still close. The model, therefore, seems to work well at all the formant frequencies of the vowel. A *formant frequency* is a frequency where energy is concentrated during the production of a speech sound. Acoustically, a formant frequency would be a vocal tract resonance during the production of a particular speech sound.

Next, let us see if our model works for the vowel /a/ (as in the word "father"). We said above that in producing /a/, the mouth is wide open and approximates a free boundary. So the boundary conditions for the vowel /a/ may be estimated as being free at each end of the vocal tract. For symmetrical systems, the wavelength of the first vibration mode is twice the length of the system. Again estimating that the vocal tract is about 24 cm long, the wavelength of the fundamental frequency of the vowel /a/ should be:

$$\lambda = 2 \times 24 \text{ cm} = 48 \text{ cm}$$

The wavelength of the symmetrical vocal tract phonating the vowel /a/ is about 48 cm or about 0.48 m. Next, we can calculate the fundamental resonance frequency of the vocal tract phonating /a/ as we did above:

$$f = c/\lambda$$

or

$$f = (343 \text{ m per sec})/(0.48 \text{ m}) = 715 \text{ per sec} .$$

So, the symmetrical vocal tract phonating the vowel /a/ should have a fundamental resonance frequency of about 715 hertz, double that of the unsymmetrical vowel /u/. Measurements of the vowel /a/ show a fundamental frequency of about 730 Hz, so our model is still working well.

Finally, we will look at the overtones of the vocal tract when producing the vowel /a/. Overtones of a symmetrical air column are whole number multiples of the fundamental frequency. The second resonance (first overtone) frequency would be:

$$f_2 = 2\,(715 \text{ Hz}) = 1430 \text{ Hz} .$$

Laboratory measurements reveal a second formant of near 1200 Hz. for the vowel /a/. The third resonance (second overtone) frequency would be:

$$f_3 = 3\,(715 \text{ Hz}) = 2145 \text{ Hz}$$

Laboratory measurements of the second overtone for phonating the vowel /a/ are around around 2300 Hz. So, our model seems to work fairly well at predicting the formant frequencies of vowels for the two specific vocal tract configurations we have examined.

So far we have looked at two very different speech sounds, /u/ and /a/. These two differed basically in the boundary condition of the mouth during phonation. In addition to changing the boundary condition at the mouth, we can alter the vocal tract by internally constricting it. Let us consider how this constriction method alters the vocal tract's output. We will look at the vowel /i/ (as in the work "sit").

When /i/ is produced, the jaw is raised and the mouth is pulled into a slitlike, almost closed position. We can model this position as a nearly fixed boundary, the way we did with /u/. Because the mouth is not closed as much as it was with /u/, the /i/ lip position really causes the node to occur in front of the mouth, outside the body, not right at the mouth. This can be taken into account by adding an *end correction* (an amount added to the actual length of an air column to account for nodes occurring outside of the system). The usual end correction is about the same as the air column's diameter. For the mouth this would be about 4 cm, giving an effective vocal tract length of 28 cm for the vowel /i/.

In addition, when /i/ is produced, the tongue is raised to the roof of the mouth at the alveolar ridge forming a constriction within the oral cavity. The alveolar ridge is quite near the lips when the lips are retracted for the /i/ position. This constriction is so near the lips that it will not interfere with the fundamental resonance frequency of the vocal tract during a near-closed lip position. By contrast, the constriction that is created by the tongue's position does interfere with antinodes of the first and second overtones of the vocal tract in this operating condition. When a constriction of an air column interferes with antinodes of certain resonance frequencies, those antinodes are suppressed and the resonances do not occur. So the usual first and second overtones of the unsymmetrical vocal tract are suppressed when the vowel /i/ is produced. The third overtone of the air column would normally have a node at the position of the constriction, however, and would not be suppressed. So the vowel /i/ should have formants at the fundamental and third overtone frequencies of the unsymmetrical vocal tract.

Let us calculate these frequencies and see how they compare with the measured formants of this vowel. For unsymmetrical system, the wavelength of the fundamental frequency is four times the length of the system. If the vocal tract has an effective length of 28 cm, then the wavelength of the first mode of /i/ should be:

$$\lambda = 4 \times 28 \text{ cm} = 112 \text{ cm} .$$

So the wavelength of the unsymmetrical vocal tract phonating the vowel /i/ should be about 112 cm or 1.12 m. From this and the speed of sound in air (343 m per sec), we can calculate the fundamental resonance frequency of the unsymmetrical vocal tract producing /i/:

$$f = (343 \text{ m per sec})/(1.12 \text{ m}) = 306 \text{ per sec} .$$

The unsymmetrical vocal tract phonating the vowel /i/ results in a fundamental frequency of around 300 Hz.

Now, let us look at the third overtone of our unsymmetrical vocal tract producing the vowel /i/ (remember, the first and second overtones have been suppressed by the tongue). Recall that the overtones of an unsymmetrical system are given by the formula:

$$f_n = (2n - 1)(f_1) .$$

So, the third overtone (fourth resonance) is:

$$\begin{aligned} f_4 &= [(2 \times 4) - 1]\,(306\ \text{Hz}) \\ &= (8 - 1)\,(306\ \text{Hz}) = 7 \times 306\ \text{Hz} = 2142\ \text{Hz} \end{aligned}$$

Laboratory measurements of the second formant frequency of the vowel /i/ show values around 2300 Hz. So our model even seems to account for constrictions within the vocal tract.

Before going on, let us summarize what we have said about the acoustical operation of the vocal tract:

a. The vocal tract operates essentially as an air column.
b. The laryngeal end of the vocal tract must be open when it is excited so it acts as a free boundary.
c. The mouth end of the vocal tract can be varied between near-closed and open extremes and can be modeled as a free or fixed boundary, or near-fixed boundary with an end correction.
d. Moving the tongue within the oral cavity can constrict the vocal tract suppressing certain antinodes, preventing some overtones from occurring.

If we had the space, the principles we have demonstrated here could be applied to calculating all of the phonated sounds of speech. These examples, however, should give you an idea of how the vocal tract operates acoustically. Notice that we have only mentioned phonated sounds like the vowels and *glides* (vowel-like consonants like /l/, /r/, and /w/). These are the sound produced by the resonance phenomena of the vocal tract operating as an air column.

ALTERNATE SPEECH SOUND SOURCES

Up to now, the only speech sound source we have considered is the larynx. We saw that the larynx acts as a siren to produce a buzzing sound at the glottis. In addition, we examined the resonance effects that take the laryngeal siren sound and amplify certain of its frequencies to produce specific speech sounds. While it is true that this is a primary sound source for speech production, the structures of the mouth and oral cavity can act as sound sources also. There are several categories of sound that can be produced at the oral cavity and mouth. Let us look at some of these.

A *plosive* is a type of speech sound that occurs when air pressure is built up behind some constriction in the oral cavity or mouth and is suddenly released. The released air suddenly impacts on the motionless air in front of the constriction and a sound occurs. This mechanism is similar to jet noise production, except that the released air comes in a single burst rather than in a sustained stream.

Let us look at the different plosive sounds we can create by releasing the air burst from different places in the oral cavity and mouth. If the constriction is created at the mouth by releasing the air from the lips, the resulting sound is a /p/ (as in the word "par"). If the constriction is at the alveolar ridge and is released by the tongue, a /t/ (as in "tar") results. If the constriction is at the oropharynx and is released by the soft palate, the sound will be /k/ (as in "car"). In these three sounds (i.e., /p/, /t/, and /k/), the laryngeal siren is not operating. These are the voiceless plosives. The voiced plosives occur if the laryngeal siren is operating. The voiced plosives are /b/ (as in "bay"; cognate of /p/), /d/ (as in "day"; cognate of /t/), and /g/ (as in "gay"; cognate of /k/).

By contrast with the plosive sound, the *stop* is a type of speech sound that is created by suddenly constricting the vocal tract, thereby stopping the speech sound. It is the opposite of a plosive sound. If the constriction is created at the mouth by the lips, the resulting sound is a /p/ (as in the word "lip"). If the constriction is at the alveolar ridge and is released by the tongue, a /t/ (as in "lit") results. If the constriction is at the oropharynx and is released by the soft palate, the sound will be /k/ (as in "lick"). As with the voiceless plosive sounds, the laryngeal siren is not operating when the voiceless stop occurs.

Voiced stops are created by continuing phonation until the stop is completed. The voiced stops are /b/ (as in "rib"), /d/ (as in "rid"), and /g/ (as in "rig"). Since the corresponding stops and plosives are produced at the same place and correspond to the same written symbols, they are often called *stop-plosives*. Keep in mind, however, that the manner of producing stops differs from that for plosives.

A *fricative* (or *sibilant*) is a type of speech sound that occurs when a jet noise is created by forcing air through a small opening in the mouth or oral cavity. When the opening is between the lower lip and upper teeth, the sound is a voiceless /f/ (as in the word "fat"). The voiced cognate of /f/ is /v/ (as in "vat").

SUMMARY

In this chapter, we have discussed the siren-like nature of phonation at the larynx. We have reviewed the acoustics of vocal fold operation using a string model and considered such phenomena as the Bernoulli effect. Finally, we have examined the various modes of phonation, including voiced, voiceless, whispered and breathy speech as well as vocal fry and glottal stops.

The process of producing speech sounds by the movement of structures in the mouth and oral cavity is called *articulation*. The *articulators* are the structures of the mouth and oral cavity that operate to produce speech sounds. The articulators we have examined in this section, the lips, teeth, tongue, alveolar ridge, and soft palate, are especially important in producing plosive, stop, and sibilant sounds. These sounds along with the nasals, vowels, glides, and glottal stop make up all the usual sounds of English. Other common English sounds are combinations of these sounds.

A glossary of phonological terms can be found in Appendix C.

CHAPTER 7
HEARING

In this chapter we will be examining the mechanical part of the hearing mechanism. The *mechanical ear* is that part of the ear that responds to sound exclusively through vibration (i.e., with a mechanical response). Figure 7.1 shows the components of the mechanical ear. The mechanical ear is made up of the *outer ear* (that part of the ear including the pinna and the external ear canal, extending medially to the eardrum membrane) and the *middle ear* (a cavity in the temporal bone medial to the eardrum membrane and containing the hammer, anvil and stirrup). In the sections that follow, we will carefully examine the structure, function, and acoustics of the mechanical ear.

THE OUTER EAR

The outer ear includes the pinna, the external ear canal and the eardrum membrane. Let us begin by looking at the pinna (see Figure 7.2).

The Pinna

The *pinna* (also called *auricle*) is a flap of skin over cartilage that projects from the side of the head around the entrance to the external ear canal. It is the visible part of the ear. Its many grooves and ridges are not there by accident. A careful examination reveals that no two pathways to the ear canal entrance cross over the same pattern of ridges and grooves. That is important in understanding one of the functions of the pinna.

The pinna is primarily helpful in determining whether a sound came from overhead, below, in front of you, or behind you. This ability is called *vertical plane localization.* It comes from listening to a sound and determining how diffraction at the pinna has changed the spectrum (frequency pattern) of the sound. Because this ability is based on guessing how diffraction has altered a sound, it works best when sounds

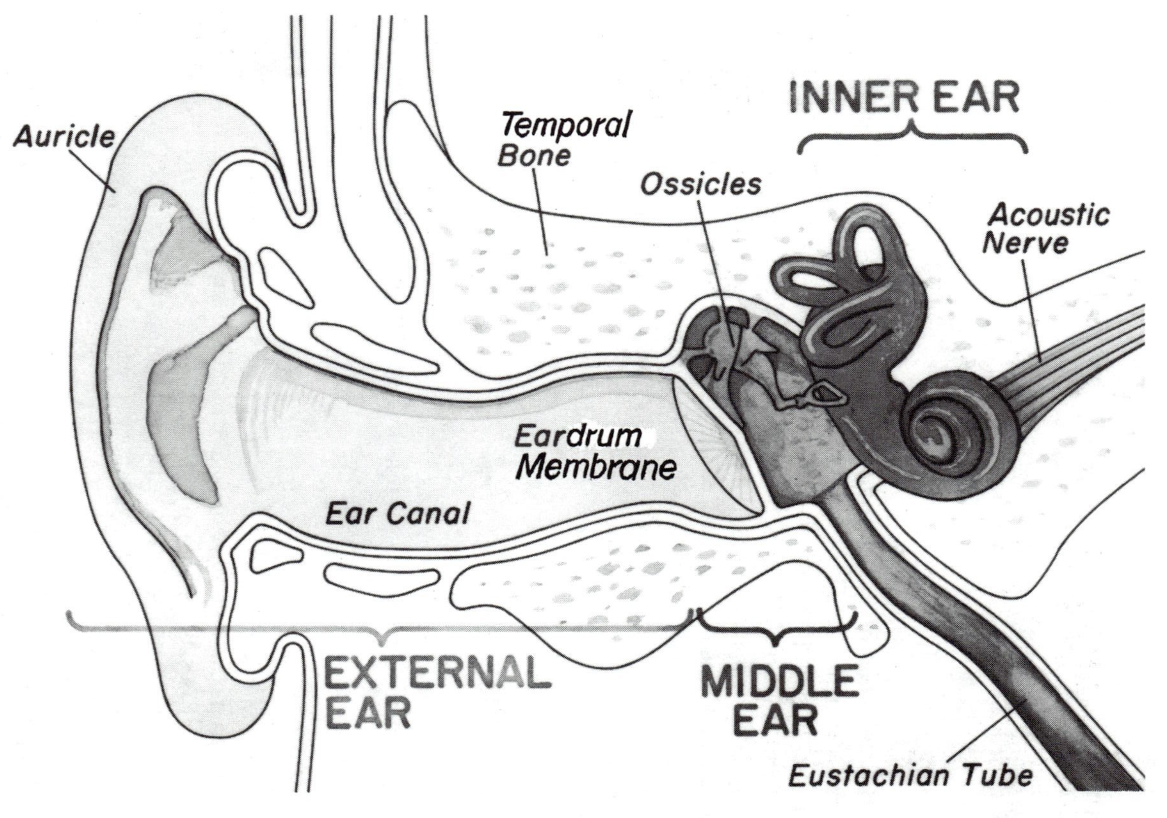

Figure 7.1. The hearing mechanism

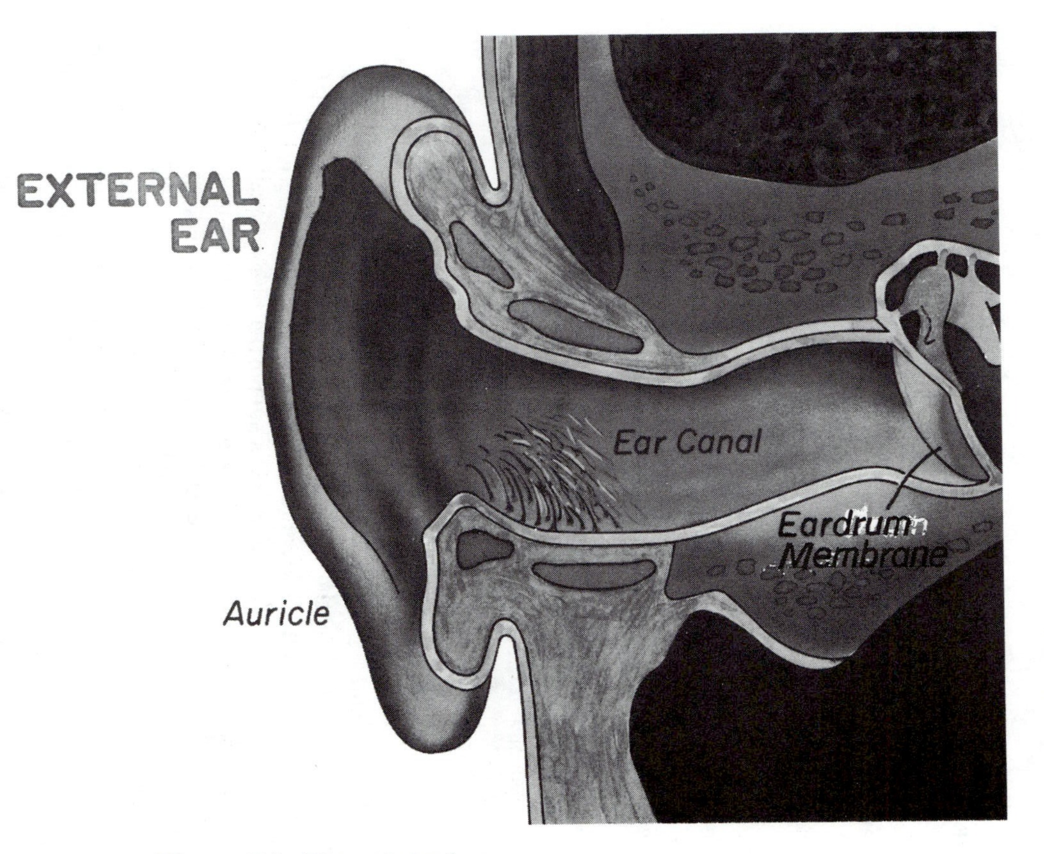

Figure 7.2. The external ear.

are familiar. People are generally much better at vertical plane localization for familiar sounds.

A second important function of the pinna is collecting sound. Looking closely at the pinna, we notice that it is generally funnel-shaped. This means that sound is collected over a large area (over the whole pinna) and delivered to a smaller area (the ear canal entrance). Remember from Chapter 4 that the amount of sound reaching a place depends on the amount of sound energy present and the area over which it is spread. If we collect sound energy over a large area and channel it into a small area, the intensity (sound energy per unit area) will increase. The pinna increases the sound energy entering the ear canal entrance by collecting it over a large area.

You can experiment with the effect of the pinna on sound loudness by enlarging the effective area of your own pinna. Cup your hand behind your ear while listening to music or a newscast. By removing your hand and replacing it several times, you will hear the increased loudness caused by collecting sound over a large area.

The Outer Ear: External Ear Canal

The *external ear canal* (*external auditory meatus*) is a tube that enters the head from the area of the pinna and extends through the skull to the eardrum membrane. (see Figure 7.2). The outer (lateral) two-thirds of the external ear canal is supported by cartilage (called the *cartilaginous portion*), while the inner (medial) third of it is supported by the temporal bone (called the *bony portion*). The cartilaginous support gives flexibility to the outer ear system so that it is less likely to be damaged by injury. You can test this by grabbing your pinna and pulling it in several directions. It moves freely because it is supported by cartilage. The external ear canal is about 1 cm in diameter and 2.5 cm long with two bends in it.

The main purpose of the external ear canal seems to be to protect the delicate parts of the middle and inner ear from foreign matter. The bends make it very unlikely that foreign matter will travel through the ear canal to its medial end. In addition, the outer two-thirds of the external ear canal contain *wax glands* (or *sebaceous glands*). These glands secrete a soft wax (called *cerumen*) that traps foreign matter. Normally, the wax glands are angled so that they project wax outward along the external ear canal. In this way, the cerumen naturally carries foreign matter out of the ear canal. Near the wax glands of the external ear canal there are usually hairs. These hairs also tend to prevent foreign matter from moving towards the middle ear.

While some people do have problems with overproduction of cerumen that can block the external ear canal, some amount of wax is normal. Somehow, people seem to be embarrassed at having cerumen in their ear canals and try to remove it with cotton tipped swabs. Do *not* do this! Cotton swabs are for cleaning the pinna, not the external ear canal. You can seriously damage your middle ear if you slip just a little while you have a cotton swab in your external ear canal. Besides, poking around in your external ear canal triggers the wax glands to produce extra cerumen. It does not help the problem. If anything, it may make matters worse.

The external ear canal also protects the structures of the middle and inner ear from injury. Because the ear canals are so long the delicate structures of the middle and inner ears are recessed well into the head. To get an idea of how far inside the head the inner ears are recessed, look at yourself in the mirror. Your inner ears are behind and a little lower than your eyes. That is pretty far into the head! Blows to the head or other surface injuries of the head are not likely to damage the delicate structures of the middle and inner ears.

In addition to its protective functions, the external ear canal has some important acoustical properties. It operates essentially as an air column. The lateral end of the external ear canal is open, while its medial end is covered by the eardrum membrane. So the external ear canal is an unsymmetrical air column that is about 2.5 cm long. In Chapter 5 we said that for unsymmetrical systems the wavelength of the fundamental frequency is four times the length of the system. If the external ear canal is about 2.5 cm long, then the wavelength of the first mode of the external ear canal should be:

$$\lambda = 4 \times 2.5 \text{ cm} = 10.00 \text{ cm} .$$

So the wavelength of the fundamental resonance frequency of the unsymmetrical external ear canal is 10 cm or about 0.10 m. From this basic value and the speed of sound in air (343 m per sec from Chapter 4) we can calculate the fundamental resonance frequency of the ear canal using the wavelength formula from Chapter 5. Recall the formula:

$$f = c/\lambda ,$$

where:

λ = wavelength
c = speed of sound, and
f = frequency .

This would result in a fundamental resonance frequency of:

$$f = (343 \text{ m per sec}) / (0.10 \text{ m}) = 3430 \text{ per sec} .$$

The external ear canal should have a fundamental resonance frequency of about 3430 hertz. Measurements of adult external ear canal resonances are between 3000 Hz and 3600 Hz, so our model is close.

The Outer Ear: Eardrum Membrane

The *eardrum membrane* (or *tympanic membrane*) is a cone-shaped membrane that seals the inner end of the external ear canal.

About five-sixths of the eardrum membrane is stretched tightly and is called the *tense part* (or *pars tensa*). The remaining one-sixth of the membrane is loose and is called the *flaccid part* (or *pars flaccida*, also called *Shrapnell's membrane*). This flaccid part is free to move when the ear undergoes sudden pressure changes, as in an airplane. Without a flaccid part, your eardrum might rupture during even minor pressure changes.

Since the eardrum membrane is a membrane, we can determine its acoustical properties as we would with any simple circular membrane. A circular membrane has some curious patterns of vibration at its various modes. Its nodal lines may be circular (running parallel to the edges of the membrane) or straight (running across the membrane and through its center as a diameter of the circle). The simplest mode of circular membrane vibration is a pattern with a single, circular node around the edges and an antinode at the center of the membrane. This is also the fundamental mode of vibration for the eardrum membrane. Only the tense part of the eardrum membrane vibrates, however, because the flaccid part has no restoring force. A calculation of the resonance frequencies of the fundamental mode of the eardrum membrane shows a fundamental resonance frequency at about 4000 Hz.

THE MIDDLE EAR

Medial to the eardrum membrane, the temporal bone opens up into the middle ear, a small mucous-lined chamber between the outer and inner ears that contains the ossicles. The *ossicles* are three small bones (malleus, incus, and stapes) in the middle ear that transfer the vibration of the eardrum membrane to the inner ear (see Figure 7.3).

You have probably noticed that we never call the eardrum membrane just "the eardrum," though it is often called that in everyday

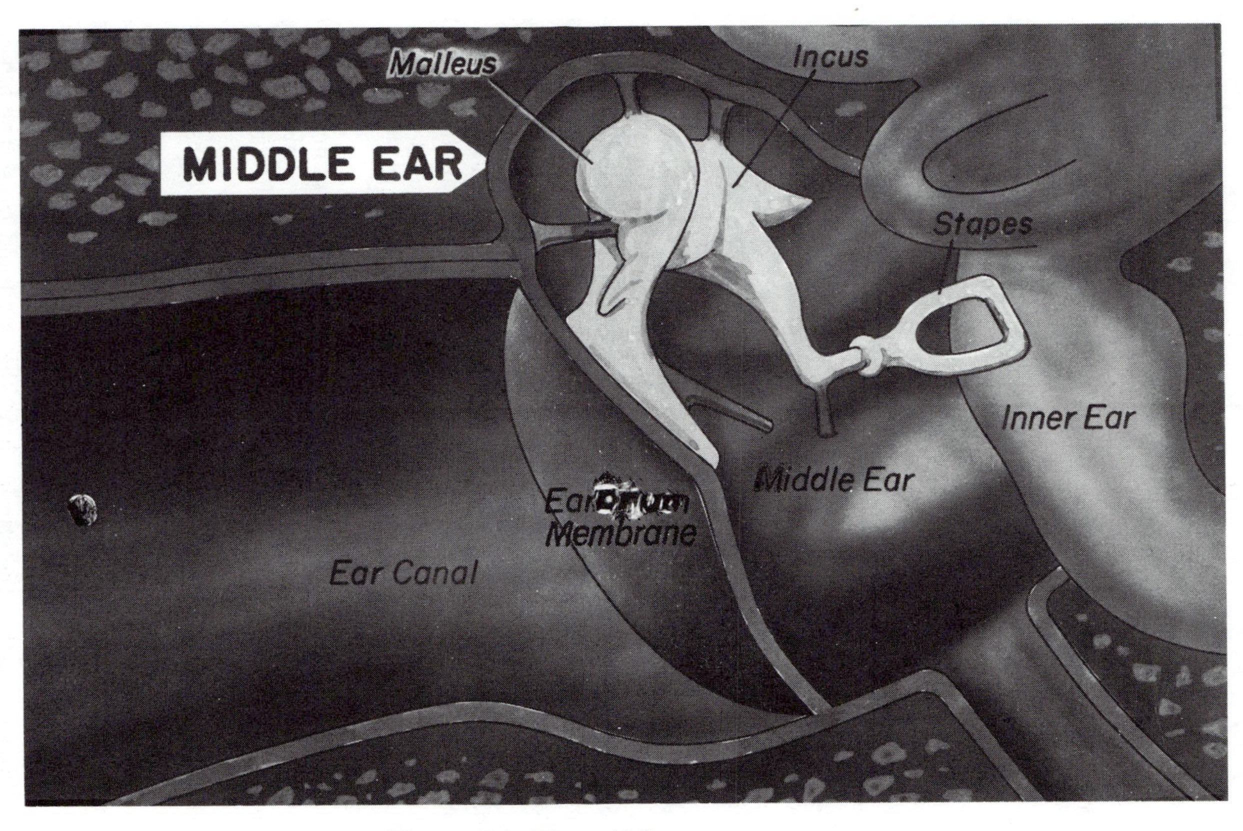

Figure 7.3. The middle ear.

speech. That is because, technically, "eardrum" does not mean the same as "eardrum membrane." The *eardrum* (also called *tympanum* which is Latin for "drum") is the eardrum membrane and the middle ear space over which it is stretched, just as the word "drum" refers to a drumhead membrane and the drum frame over which it is stretched.

The Middle Ear Space

The middle ear space is lined with a mucous membrane that breathes oxygen out of the air in the middle ear space. Because of this, it is necessary to exchange the air in the middle ear space periodically or its mucous lining will die. The *Eustachian tube* (see Figure 7.1) is a tube extending medially and inferiorly through the temporal bone from the middle ear space to the nasopharynx. When it is opened, the eustachian tube allows the exchange of air from the middle ear space with that in the nasopharynx, replenishing the oxygen content of the middle ear.

The muscles of swallowing serve to open the Eustachian tube. That is why chewing gum relieves pressure sensations at the ear during flying. The *ossicular chain* hangs freely in the middle ear space, suspended by ligaments. It delivers sound-induced vibrations from the eardrum membrane across the middle ear space to the *oval window* where sound waves enter the inner ear.

The *hammer* or *malleus* (see Fig. 7.3) is the most lateral of the ossicles and is attached to the eardrum membrane and the anvil. Vibrations of the eardrum membrane in response to sound in the external ear canal will be transferred to the hammer through its handle and from there to the body of the hammer where they are transferred to the anvil.

The *anvil* or *incus* is the middle bone of the ossicular chain, attached laterally to the hammer and medially to the stirrup. It is also divided into two major parts. The body of the anvil is rigidly attached to the hammer at the *malleo-incudal joint* and receives vibrations from the hammer through this joint. These vibrations are then transmitted to the stirrup.

The *stirrup* or *stapes* is the most medial of the ossicles and is attached laterally to the anvil at the *incudo-stapedial joint*. This joint is not rigid, but allows a free, pivoting movement of the stirrup relative to the anvil.

Vibrations enter the stirrup from the anvil and cause the stirrup to slide into and out of the oval window like a piston in a cylinder. This delivers the vibration to the fluid of the inner ear.

Two muscles are attached to the ossicular chain. These *middle ear muscles* are triggered by loud sounds and reduce the amplitude of vibrations before they enter the inner ear. The one drawback of the system is that a loud sound must reach the inner ear to trigger this *acoustic reflex*, so, for brief loud sounds, "the coop door is closed after the weasel gets in."

During normal response to sound, the hammer and anvil vibrate transversely as a single unit, while the stirrup vibrates longitudinally as a separate element of the ossicular chain. Resonance frequencies of bars in transverse motion are generally very high and, for the ossicles, they are higher than the highest frequencies humans can hear. That means they cannot affect human hearing, so we will not try to calculate the transverse resonance frequencies of the hammer and anvil. Even though a system is primarily vibrating transversely, it will have some longitudinal motion at the same time. So the longitudinal resonance frequency of the combined hammer-anvil can affect hearing and we will determine its resonances. Likewise, the resonance of the stirrup in longitudinal vibration can affect how we hear, so we will examine that also.

Though they obviously have intricate shapes, the ossicles can be treated as simple bars when we model them acoustically. Remember that longitudinal vibration of a bar follows the same pattern as longitudinal vibration of an air column. So the formulas for calculating the resonance and overtone frequencies of ossicle vibration will be the same as those for air columns. Table 7.1 summarizes the various resonances of the components of the mechanical ear.

These resonances have a significant effect on hearing. Notice that most of them occur in the range from about 400 Hz to about 6000 Hz,

Table 7.1. Resonant Frequencies of Mechanical Ear Structures

Structures	*Resonances (in hertz)*		
External Ear Canal	f_1: 3440	f_2: 10320	f_3: 17200
Eardrum Membrane	$f_{0,1}$: 4000 $f_{0,2}$: 9200	$f_{1,1}$: 6400 $f_{1,2}$: 11600	$f_{2,1}$: 8400 $f_{2,2}$: 14000
Hammer-Anvil	f_1: 402 f_4: 1608	f_2: 804	f_3: 1206
Stirrup	f_1: 1340	f_2: 2680	f_3: 4020

with only the higher overtones of the eardrum membrane and the external ear canal having resonances above 6000 Hz. In fact, this frequency region (400 Hz to 6000 Hz) is the range where hearing is most sensitive. Because of these resonances, sounds delivered to the ear near a resonance frequency will be amplified (made louder) before being delivered to the inner ear. These more intense sounds are more easily heard, so we are more sensitive to sounds near the resonance frequencies of the mechanical ear system.

Did you notice that there are two resonances near 4000 Hz (the fundamental of the eardrum membrane and the second overtone of the stirrup)? These resonances will interact with each other to produce more amplification near 4000 Hz than at other frequencies. A similar effect occurs around 1500 Hz where three mechanical resonances are near each other (the fundamental of the stirrup and the second and third overtones of the hammer-anvil). These also interact to amplify sound more than at other frequencies. In fact, human listeners are most sensitive to sound around 1500 Hz. They are also especially sensitive to sounds at 4000 Hz. So differences in human sensitivity to sound at various frequencies depend on the resonances of the mechanical ear components.

The Middle Ear: Impedance Transformer

Before leaving our study of the middle ear, we will study how it works as an impedance transformer. An *impedance transformer* is any system that bridges the difference between two media having different characteristic impedances to allow a wave to be transmitted from one medium to another. The middle ear acts as an impedance transformer between air and inner ear fluid.

Three factors contribute to the impedance transformer action of the middle ear:

1. The area ratio of the eardrum membrane to the stirrup footplate
2. The lever ratio of the hammer to the anvil
3. The buckling effect of the eardrum membrane

Let us see how each of these works.

The *area ratio* for the middle ear is the area of the eardrum membrane divided by the area of the stirrup footplate. The area of the eardrum membrane is about 35 times that of the stirrup footplate. All the force collected at the large eardrum membrane is delivered at the smaller stirrup footplate. The pressure at the stirrup footplate is 35 times that at the eardrum because pressure is force divided by area. The principle involved here is like that for a thumbtack. By collecting the force over a large thumbtack head and delivering it over the small point, you can push the thumbtack into a piece of wood. Using the same force, you could not push your thumb through the wood. Why? Pressure! All the force delivered to the head of the thumbtack is concentrated on that small point that touches the wood. The pressure is much larger for the thumbtack point than for its head. Similarly, the sound pressure is 35 times larger at the stirrup footplate than it was at the eardrum membrane.

The *lever ratio* for the middle ear is the length of the hammer divided by the length of the anvil. The hammer is about 1.5 times longer than the anvil and they are fused, so they move as one bone. In fact, their combined shape is similar to that of a crowbar or prybar and the prybar is a good model of how they work. In any lever, a force delivered to the long arm pushing it a long distance is converted to a larger force delivered to the short arm moving it a short distance. If you pull out a nail with a prybar you move the handle a long way with a small force, yet it pulls the nail a short way out with a much stronger force. Exerting the same force with your bare hands, you would never be able to pull the nail out. By using the lever, you are trading less distance moved for more force delivered. The same thing happens in the middle ear. The tip of the anvil moves only 2/3 as far as the hammer, but the anvil moves with 1.5 times the force.

The *buckling effect* for the middle ear is the force increase caused by the bending or "buckling" of the eardrum membrane as it moves in response to sound. The buckling effect is similar to the leverage effect described above. Because the eardrum membrane is conical, it buckles, moving more at its center that at its edges. This means that the hammer moves only about half as far as the surface of the eardrum membrane. This results in the same kind of trading effect we described for the lever above. By moving half as far, the hammer actually moves with about two times the force of the eardrum membrane.

These three factors interact to produce the *middle ear transformer effect*. In each of these cases, the pressure is increased by the middle ear. The area ratio increases the pressure by delivering the eardrum force over a footplate area that is 1/35th the size of the eardrum membrane. In this case, the force is the same but the area is changed to increase the pressure. For the lever ratio and the buckling effect, less distance moved is traded for a direct increase in force. Here pressure increases because force increases. The overall increase in pressure from the sound entering the ear to the sound reaching the inner ear is the product of these three factors (the area ratio of 35, the lever ratio of 1.5, and the buckling effect of 2):

$$35 \times 1.5 \times 2 = 105$$

So the sound reaching the inner ear through the middle ear mechanism is about 100 times that which enters the external ear canal.

Let us see how this affects the impedance difference between air and inner ear fluid. When a sound wave of 50 dB sound intensity level (i.e., 0.0000001 watt per square meter) enters the external ear canal, only about 29 dB of sound intensity level (i.e., 0.0000000008 watt per square meter) enters the inner ear. Because of the impedance difference between air and inner ear fluid, only about 1/125th of sound energy can be transmitted from one to the other. The middle ear impedance transformer increases the sound reaching the inner ear 105 times. Thus, the energy reaching the inner ear is about $105 \times 1/125 = .84$ of the energy entering the ear canal.

Specifically, 21 dB is lost by reflection because of the impedance difference between air and inner ear fluid, while 20 dB is gained because of the middle ear impedance transformer. This combined effect of the impedance difference and the impedance transformer leaves us with a net loss of only 1 dB of sound intensity level.

THE INNER EAR

The Cochlea

In this section we will look at transduction. *transduction* is the process of converting energy from one form to another. The *cochlea* is a structure within the temporal bone whose primary function is transducing sound waves (delivered to it by the mechanical ear) into neural signals. This transduction process is carried out by an intricate and delicate mechanism especially designed to convert sound into nerve impulses that carry a code for the fine details of information present in sound waves.

For a minute, picture a snail creeping along a thick blade of grass. Think carefully about its shell. This spiral shape is common in nature and we will be thinking about it a lot in this chapter. The cochlea is inside a spiral-shaped tunnel in the temporal bone, just medial to the middle ear space. Note from Figure 7.4 that the tunnels for the semicircular canals of the balance mechanism are attached to the spiralling cochlea. The continuous set of tunnels in the temporal bone that

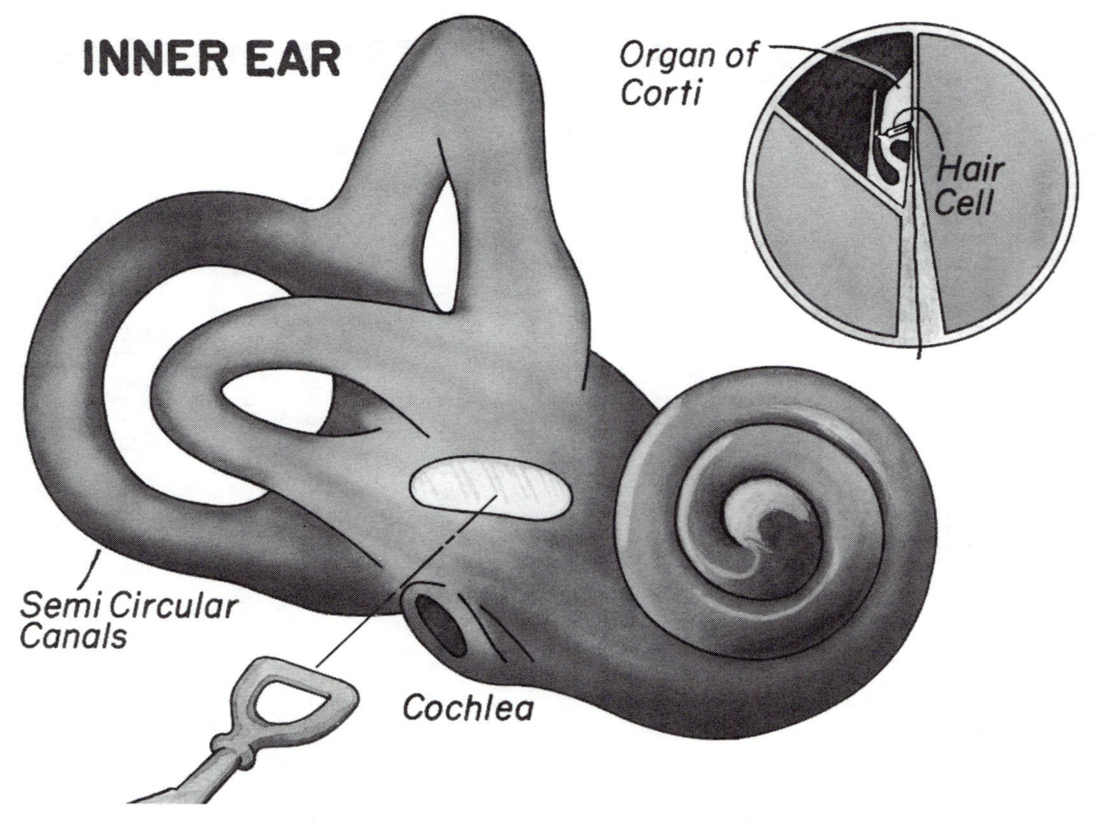

Figure 7.4. The inner ear.

contain the cochlea and the semicircular canals are called the *labyrinth*. The Labyrinth was the confusing set of caves that were the home of the Minotaur, a mythological Greek beast. Apparently, the first anatomists to discover the cochlea and the semicircular canals thought they were a confusing set of caves in the temporal bone and named them accordingly.

It is important to understand that, in the human, the labyrinth is a tunnel in the bone. I do not want to give the impression that the cochlea is a snail-shell-shaped piece of bone that can be removed intact from the head. The spiral opening is just a space in the temporal bone and it is the shape of that space that reminds us of a snail's shell. In the center of this open space is the *modiolus*, a column of bone that forms the center post of the cochlea's labyrinth. The modiolus is filled with tiny tubes through which auditory nerve fibers and blood vessels pass so that they can reach every part of the cochlea.

The three semicircular canals are arranged roughly at right angles to one another (see Figure 7.4). They are the *superior semicircular canal* which is in the vertical plane and extends front to back in the head, the *posterior semicircular canal* which is in the vertical plane and extends left to right in the head, and the *lateral semicircular canal* which is in the horizontal plane.

Picture a boxer who ducks a punch by suddenly twisting his head up, back and to the side. You can visualize this because any motion can be described by its component motions in three dimensions ("up," "back," and "to the side"). Any movement of the head can be represented by its component motions in the upward (superior semicircular canal), backward (posterior semicircular canal), and sideways (lateral semicircular canal) directions. That is why the three semicircular canals are arranged the way they are.

The cochlea is a long, soft tissue tube that lines the spiral-shaped part of the inner ear. That end of the cochlea nearest the middle ear is the *base* (or *basal turn*), while the *apex* (or *apical turn*) is the end farthest from the middle ear (see Figure 7.4). The cochlea is divided into three ducts, as shown in Figures 7.4 and 7.5.

One of the three ducts of the cochlea, the *scala vestibuli* begins at the oval window in the basal turn and extends to the apex. Because sound reaches the inner ear through the oval window, that part of the inner ear nearest the oval window is called the *vestibule* meaning "entrance way".

Another of the three ducts of the cochlea is the *scala tympani* which is adjacent to the scala vestibuli and extends from the base of the cochlea to its apex. This duct opens directly into the middle ear

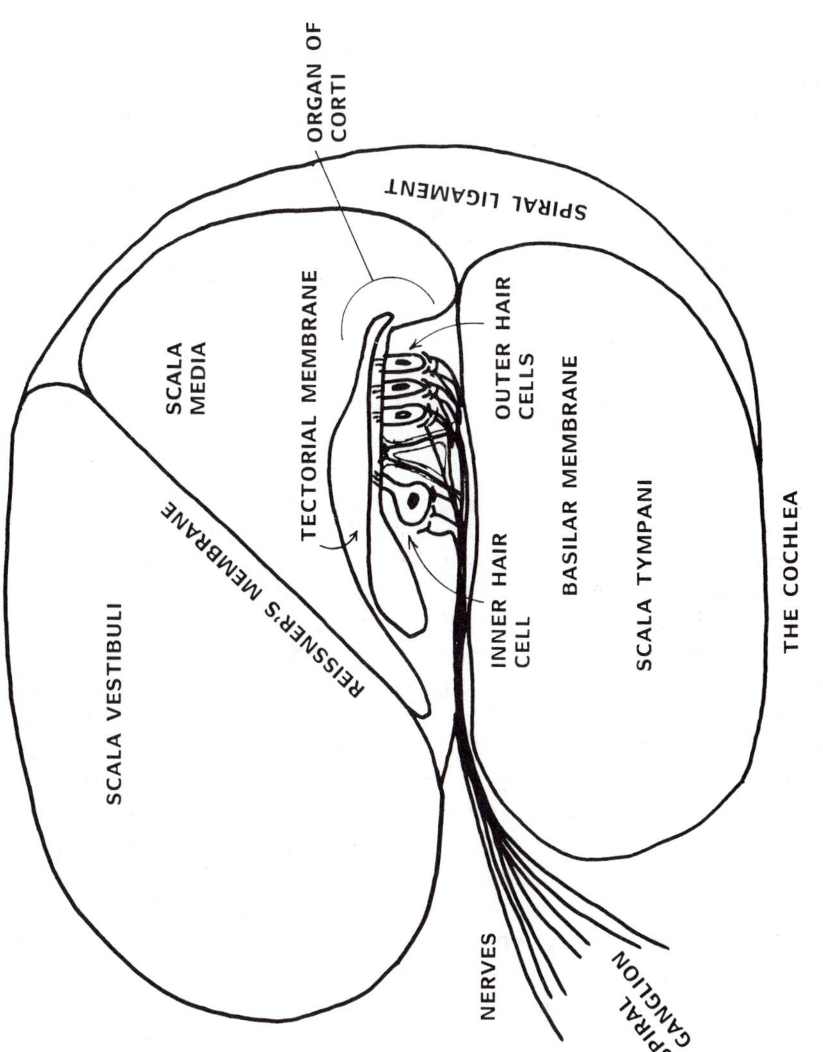

Figure 7.5. Cross-section of the cochlea.

space (or eardrum) through the round window, so its name means "the duct of the drum". These two ducts are connected to one another at the apex by an opening called the *helicotrema.* So both the scala vesibuli and the scala tympani are filled with the same fluid, *perilymph.*

Between these two large ducts of the cochlea there is another, narrower duct called the *scala media,* which means "the duct in the middle." The scala media is filled with *endolymph,* an ionic fluid that is separate from perilymph. Reissner's membrane is a sheet of soft tissue that separates the scala media from the scala vestibuli. The *basilar membrane* divides the scala media from the scala tympani. Before closing this section, we must identify one final structure of the scala media. The *organ of Corti* is a trapezoid-shaped structure that rests on the basilar membrane. It is here that transduction takes place.

Waves in the Labyrinth

Before examining the organ of Corti, let us review how waves travel in the membranous labyrinth of the cochlea. The model for wave motion in the cochlea is ocean waves nearing a beach. The physics of surf is almost identical to the physics of wave motion in the cochlea. Waves in the ocean always begin in the deep water and travel toward the beach. As they near the beach, they get bigger (increase in amplitude) and begin to crest or curl at the top. Eventually, the wave crests so far that it collapses upon itself or "breaks" in a large curl. Once a wave has broken, only a small ripple continues from there to the beach. If you have spent any time at the ocean trying to ride the surf, this description should sound familiar. You also should have noticed that some waves break in the deep water, while others travel almost to the beach before breaking. The trick in surfing is to catch the waves that travel near to the beach for a long ride. In this section we will examine wave motion in the cochlea. We might also pick up a few pointers in surfing!

When sound causes the stirrup to move into the oval window, it puts pressure on the perilymph of the scala vestibuli, near the oval window. Liquids are *incompressible* (cannot be made more dense than they already are) so an inward force (push) at the oval window must result in an outward force from the scala vestibuli somewhere. The force is transmitted from the scala vestibuli, through the scala media, into the scala tympani, putting pressure on the perilymph in the scala tympani. Notice that the scala media bulges downward into the scala tympani as this happens.

The force from the scala vestibuli is transmitted downward through the incompressible perilymph of the scala tympani to its bony floor where it stops because the bone will not bend. This is where the size pattern of the cochlea becomes important. It is easier to dissipate pressure in deep fluid than it is in shallow, so when the force reaches the floor of the scala tympani, it turns toward the base (deeper direction) rather than toward the apex (shallower direction). After the force turns basalwards from the scala tympani floor, it reaches the motionless perilymph near the base and the force is turned back toward the scala vestibuli. Next, the force is transmitted back through the scala media and into the scala vestibuli, causing the scala media to bulge upwards into the scala vestibuli, just basal of its downward bulge. A similar swirling pattern of force transmission happens beneath ocean swells for the same reasons described above.

In fact, the force has swirled around in an oval pattern carrying some of the perilymph with it. This swirling fluid motion is called an *eddy current* and it controls the pattern of wave movement in the cochlea. Surfers and ocean bathers call these eddy currents "undertow" and undertow controls waves nearing an ocean beach. The force swirls around once forming two bulges in the wave. These two bulges are pushed toward the apex (toward the shore at the ocean) by the force which has swirled around in the eddy current.

These moving bulges are a wave because the bulges move toward the shallow area, but the fluid (perilymph or sea water) does not. As a wave is pushed toward the shallows, new eddy currents are produced and the wave is pushed further and further along. In summary, waves tend to form at the base of the cochlea and move toward its apex because the depth difference within the cochlea creates eddy currents. Similarly, ocean waves move toward the beach because the depth differences near the shore create undertow.

In addition to pushing the wave toward the shallows, the eddy currents affect the wave in one other way. Some amount of force is lost in transmitting the force around the eddy current because of friction between the particles of the fluid. As the wave moves into the shallow area, the force is returned to the surface sooner, because the force has less distance to travel. Since the eddy currents are shorter in the shallower fluid, less force is lost due to friction. As the wave moves toward the apex (or the beach), the force pushing it gets stronger and increases the amplitude of the wave. So the amplitude of the wave increases continuously as it moves toward the apex (or beach).

As the wave's amplitude increases, the slope between the wave's two bulges gets steeper. This slope continues to steepen until it becomes vertical. At that point, the wave cannot hold itself up any longer and it collapses, with only a tiny ripple travelling the rest of the way to the apex (or beach). Remember we said at the beginning of this section that some ocean waves travel all the way to the beach while others break in the deep water. The same is true for waves in the cochlea. Some waves collapse near the base while others travel all the way to the apex. Frequency controls this effect. In both the ocean and the cochlea, high frequency waves collapse in the deep fluid, while low frequency waves travel near to the beach (or apex).

In surfing that means the best waves, the ones that give the longest rides, are the low frequency ones. In the cochlea this means that you can tell the frequency of a wave by where it crested. If the wave crests near the oval window, it is a very high frequency sound, between 10,000 Hz and 20,000 Hz. A 2000 Hz wave would crest about halfway between the base and the apex, while a 200 Hz wave would crest at the apex. The *place theory* is a hearing theory that we determine the frequency of a sound by where its wave crests within the cochlea. Scientist agree that this is one of the ways (but not the only way) the hearing mechanism determines the frequency of a sound in the cochlea.

The Organ of Corti: Shearing Forces

Within the scala media, we have described the trapezoid-shaped structure called the organ of Corti (or Corti's organ). We have said that this is the place where transduction from an acoustic wave to a neural message happens. In this section, we will look at some of the structures of Corti's organ so we can begin to see how transduction takes place. Figure 7.5 presents the organ of Corti within the scala media. It is supported by the basilar membrane and extends between the bony shelf of the modiolus and the spiral ligament. The organ of Corti has two major parts, the body of the organ of Corti which contains the sensory cells and rests on the basilar membrane, and the *tectorial membrane* which forms a roof or covering over the body of the organ of Corti.

When a sound wave travels in the cochlea, the entire scala media moves toward the scala tympani and then toward the scala vestibuli. This is an up-and-down movement in Figure 7.5. So a sound entering the cochlea will cause this up-and-down movement of both the tectorial membrane and the body of the organ of corti. Because of the support arrangements of these two structures, a very special relative motion occurs between them. To see what this motion is like, we will do an experiment.

You will need both your hands up to the elbows for this, so put this book down (keep it open nearby so you can follow the directions). Now, put your elbows on the table and place your hands together in the "praying hands" position. Next, without moving your wrists and keeping your palms vertical, move your fingers to the left and right.

Did you notice that you can do this only by sliding your fingers over one another? If your fingers are not sliding over one another, you are bending your wrists. The reason for this sliding effect is that the fingers of your two hands are hinged at different points. Your left fingers pivot from your left knuckles and your right fingers pivot from your right knuckles. Now, let us see how this applies to the organ of Corti.

Note that the tectorial membrane pivots from a different point than the body of Corti's organ (see Figure 7.5). Because of the two pivot points, the tectorial membrane slides over the body of Corti's organ just as your fingers slid over one another during our experiment. The *shearing action* of Corti's organ is this sliding of the tectorial membrane over the body of Corti's organ when sound waves enter the cochlea.

As shearing action happens within Corti's organ, a force is developed between the tectorial membrane and the body of Corti's organ. The force arising from the shearing action within the organ of Corti is called a *shearing force.* These shearing forces are essential to the process of transduction.

The Organ of Corti: Anatomy

Now that we have seen how sound waves in the cochlea lead to shearing forces in Corti's organ, let us look at the anatomy of Corti's organ in more detail. Most of the body of Corti's organ is made up of *support cells,* cells which hold up and give shape to anatomical structures. In addition to support cells, the organ of Corti contains the *hair cells,* the sensory transducers that convert sound waves into nerve impulses (see Figure 7.5). Throughout most of the cochlea, there are three rows of *outer hair cells* (long, tube-shaped sensory cells with small hairlike *cilia* that extend between the body of Corti's organ and the tectorial membrane). Each outer hair cell has about sixty cilia arranged in three V-shaped rows. The outermost row of cilia contains the tallest cilia and from each outer hair cell these cilia extend to and are embedded in the tectorial membrane. The other two rows just project into the fluid between the reticular lamina and the tectorial membrane. These cilia are very important for transduction.

Now let us look at the inner hair cells. In all parts of the cochlea, there is only one row of *inner hair cells* (thick, bag-shaped sensory cells with cilia that extend beyond the reticular lamina toward the tectorial membrane). About forty cilia project from each inner hair cell in a tuft. None of the cilia reach the tectorial membrane. They just extend into the fluid between Corti's organ and the tectorial membrane. As with the cilia of the outer hair cells, the inner hair cell cilia are essential to the process of transduction.

Recall from the last section that shearing forces arise between the tectorial membrane and the body of Corti's organ when sound waves enter the cochlea. These shearing forces will obviously bend the outer row of cilia in the outer hair cells, but what about the other cilia? Well, when shearing action happens, the fluid near the tectorial membrane will tend to move with it. This creates a current in the fluid near the tectorial membrane, a kind of cochlear breeze that bends the tops of the cilia that are free to move. These free cilia are bending in the current near the tectorial membrane. Thus, the shearing forces cause a bending of hair cell cilia. This bending of the cilia triggers the transduction process by initiating nerve impulses in the auditory nerve. These impulses lead to hearing responses in the auditory components of the brain, a topic too detailed for coverage in this text.

SUMMARY

In this chapter we have examined the mechanical hearing mechanism which includes the outer ear (pinna and external ear canal) and the middle ear (eardrum membrane, ossicles, middle ear muscles, and middle ear space). We have seen that these structures exert several acoustical effects on incoming sounds, including resonance effects which amplify certain incoming frequencies and impedance transforming effects which match the acoustical impedance of the inner ear to that of the air surrounding the head. The impedance transformer operates through the area ratio between the eardrum membrane and the stirrup footplate, the lever ratio of the hammer and anvil, and the buckling effect of the eardrum membrane.

In addition, we have examined the structure and operation of the cochlea, the labyrinthlike structure within a spiral opening in the temporal bone which contains the hair cells, the receptor cells for hearing. We have seen that the graduated depth of the cochlea causes a cresting wave (like an ocean wave) to travel from the base of the cochlea toward its apex whenever sound reaches the inner ear. These waves crest at different places along the cochlea depending on their frequency. We have learned that these waves cause shearing forces between the tectorial membrane and the reticular lamina, bending the hair cell cilia and triggering a neural response in the auditory nerve that will stimulate the auditory segments of the brain.

A glossary of audiological terms can be found in Appendix C.

CHAPTER 8
ELECTRO-ACOUSTICS

Many historically significant acoustical experiments were performed by purely mechanical processing of acoustical energy. Spectrum analysis was performed by listening to a complex sound through tuned resonators (Helmholtz resonators). Reverberation phenomena were studied by generating a very loud sound with steam-driven organ pipes and measuring the time required for this sound to die out to the threshold of hearing (Sabine). The first commercial sound recordings were made with human lung power (Caruso) "amplified" by an inverted megaphone to drive a cutting stylus attached to a diaphragm at the small end of the megaphone.

The use of electricity and electronics has widened the scope of acoustics and provided the equipment necessary to make precise acoustical measurements and to store acoustical information for analysis or regeneration at a later time. The combination of acoustics with electronics is commonly called *electro-acoustics*. It is nearly impossible today to make acoustical measurements without using electronic equipment.

ELECTRO-PHYSICS

The exact nature of electricity is difficult to describe and indeed many volumes have been written about electricity and its effects. In fact, it is easier to describe the effects of electric current than it is to define what electricity is. For the present, let us consider *electricity* to be composed of groups of particles having a peculiar property called a *charge* of electricity. It is really this charge we are concerned with and not the particle. The most common charged particle is the electron which has a charge of 4.80×10^{-10} electro-static units. No electrical charge smaller than this has ever been discovered, so it has been concluded that electricity consists of tiny discrete charges of this magnitude. The movement of these tiny discrete charges through a conducting medium is called *electric current*.

Static Electricity

Notice the word *movement* in the definition above. Movement of electrical charge produces electric current and is responsible for all of the effects that electricity is known to produce. What then is *static electricity?* The word static implies the absence of movement; static electricity is a group of these tiny discrete charges which have piled up in one place and are not moving. The charge on a cloud in a thunderstorm represents just such a static charge. When this charge builds up enough to jump between the cloud and the earth, the charge will *move* in the form of lightning which then constitutes and electric current.

The buildup of charge in the form of a group of electrons constitutes a negative charge. While it is possible to find a similar buildup of positive charge, positively charged particles are not nearly so easily moved as electrons and it is more common to find a positively charged body produced by virture of the *absence* of electrons.

Electric Current

An efficient way to produce a buildup of electrons on one body and an absence of electrons on another body is by chemical action such as that in a flashlight cell. In the common flashlight cell, chemical action causes a build-up of electrons on the outer terminal (the zinc shell) of the cell and a deficit of electrons (positive charge) on the center terminal (this is actually a carbon rod) of the cell.

When one terminal of a light bulb is connected to the outer terminal (the zinc shell) of the flashlight cell and the other terminal of the light bulb is connected to the center terminal of the flashlight cell (this happens when you switch on the flashlight), the bulb will light. As long as the external circuit is completed, that is, the conducting path from one terminal of the cell through the light bulb to the other terminal of the cell, the bulb will light until the chemicals are used up or the circuit is broken (disconnected). The electrical current in this example flows only in one direction and flows at a steady rate determined by the electrical properties of the circuit elements. The steady current in this example is called *direct current* (*dc*).

Notice in the above example that it was necessary to complete the conducting path from one terminal of the cell through the bulb to the other terminal of the cell to produce a flow of current. The current flowing through the light bulb produced useful light. A complete conducting path must always be provided from one terminal of an electric source (the cell in this example) through the electrical load (the light bulb in this example) to the other terminal for current to flow. This is commonly referred to as a *circuit*. The switch in a flashlight completes the circuit to light the bulb.

If a conductor is connected from one terminal of an electric source directly to the other terminal, a very high current will flow but little use can be made of it since all the electrical energy is converted to heat within the source. This is called a *short circuit*. In your house, this will cause a fuse to "blow" or burn out. If the conducting path is not completed, as when the flashlight switch is turned off, *no* current will flow and the circuit is called an *open circuit*.

Alternating Current

Although a flashlight cell is a convenient portable source of electricity, much energy is lost if the load (such as the light bulb) is located a long distance from the source (the cell).

An efficient way to transmit electrical power over long distances is to use alternating current. An example of this is the power supplied to our homes through long transmission lines from the generating station where rotary turbine-driven generators produce electricity.

Because the electrical current is produced by rotating machinery, the current flow can be made to change direction (and amplitude) periodically in a sinusoidal way. It is this property that accounts for the name *alternating current* (ac). A graph showing the polarity (direction of flow) and the magnitude of the current as a function of time is shown in Figure 8.1. It is easy to see that the polarity and magnitude are constantly varying and that the electrical properties of alternating current must be described in a special way. If we compare the power delivered to a load by a dc source and an ac source, we find that each delivers the same power if the root-mean-square (rms) value of the alternating current is equal to the value of direct current. The peak or maximum value of the sinusoidal alternating current is actually 1.414 times this rms value and the alternating current magnitude actually passes through zero. This alternating or varying property of alternating current produces many other useful effects which will be described

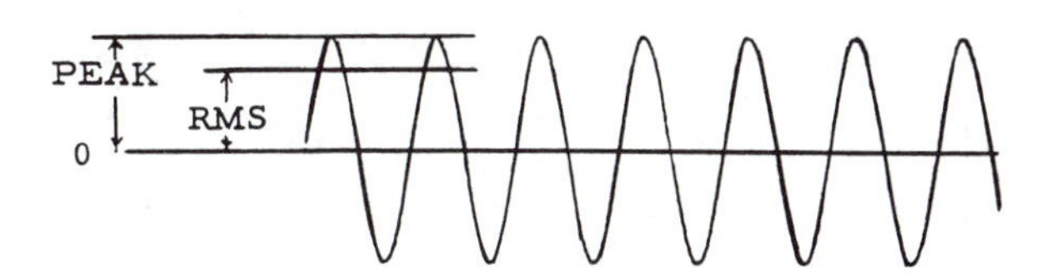

Figure 8.1. Graphical representation of alternating current.

later. Alternating current and its effects will be our principle interest in this text.

The alternating current which we receive from the power station is generated by rotating machinery but there are many other ways to generate alternating current.

The rotating machinery at the power station generates electricity by moving a coil of wire in and out of a magnetic field. This action produces alternating current which changes direction of flow and magnitude at a regular rate determined by the speed of rotation of the generator. In general, whenever a wire moves in a magnetic field, a voltage is generated in the wire. The frequency and magnitude of the voltage depends on the rate of movement relative to the magnetic field.

Simple Circuits

An electrical circuit consists of a collection of wires, switches, resistors, capacitors and inductors that perform a function whenever a current flows through them. These circuits can be represented by a drawing called a *schematic diagram* showing how the current flows from one terminal of the voltage source through the circuit elements and back to the other terminal of the source. As an example, Figure 8.2 shows a schematic diagram of a flashlight. The switch is shown in the *open* position. Under these conditions, no current can flow and the bulb does not light. When the switch is *closed*, current can flow from one terminal of the battery through the light bulb and back to the other terminal of the battery. Note that when the switch is in the closed position, the current flowing through the light bulb also flows through the switch. We say that the switch is *in series* with the bulb, likewise the bulb is in series with the switch.

If a second light bulb is connected so that it will also light when the switch is closed, it should be connected as shown in Figure 8.3. The two light bulbs are *in parallel* and the current flowing through the switch is now equal to the sum of the currents flowing through each individual

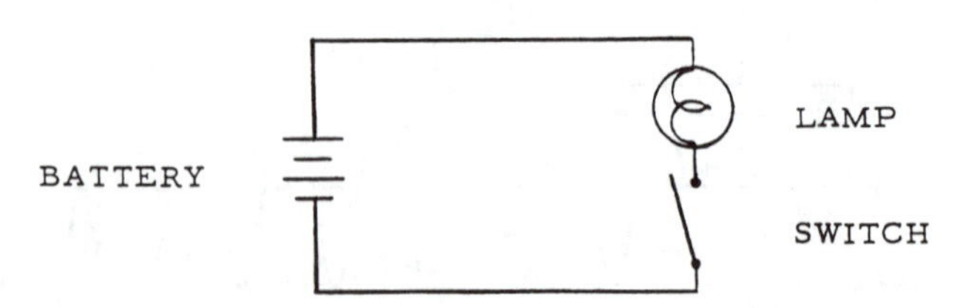

Figure 8.2. Circuit of a flashlight.

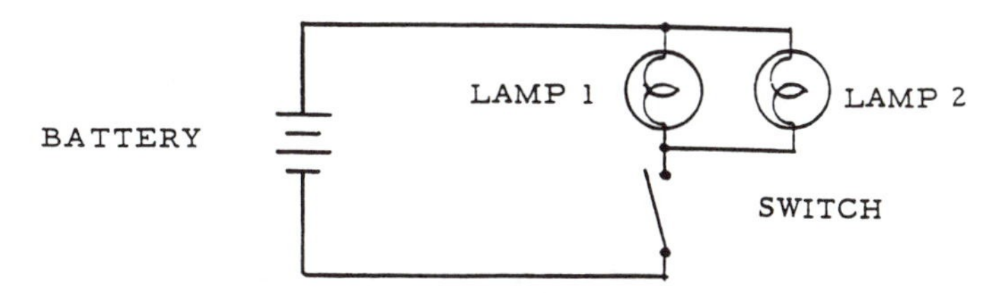

Figure 8.3. Circuit with two lamps in parallel.

light bulb. The switch is still in series with the load (the two light bulbs in parallel) and when the switch is closed the full battery voltage appears across each of the light bulbs.

Now let us consider a different circuit using the same components. In Figure 8.4 the two light bulbs are now wired in series. In this case, whatever current flows through one of the bulbs also flows through the second bulb; if one bulb should burn out (become an open circuit), the current will no longer flow, just as though the switch were open. Some sets of Christmas tree lights are wired this way. The switch and light bulbs in this circuit are all wired in series and the same current flows through each of the circuit elements. Note that the voltage across each light bulb is now *less* than the battery voltage but that the *sum* of the voltages across the light bulbs is equal to the battery voltage. (We assume that there is no voltage across the switch when it is in the closed position.)

Another variation on this same circuit is shown in Figure 8.5. The light bulbs and the switch have each been replaced by a circuit element called a *resistor*. As in the previous circuit, the same current flows through each of the resistors. The voltage across the resistors in series

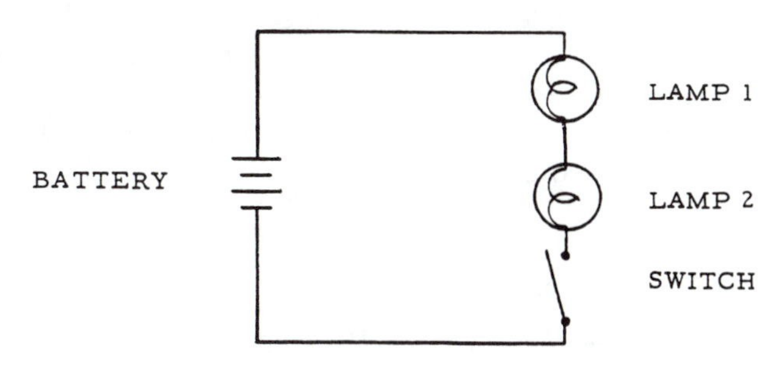

Figure 8.4. Circuit with two lamps in series.

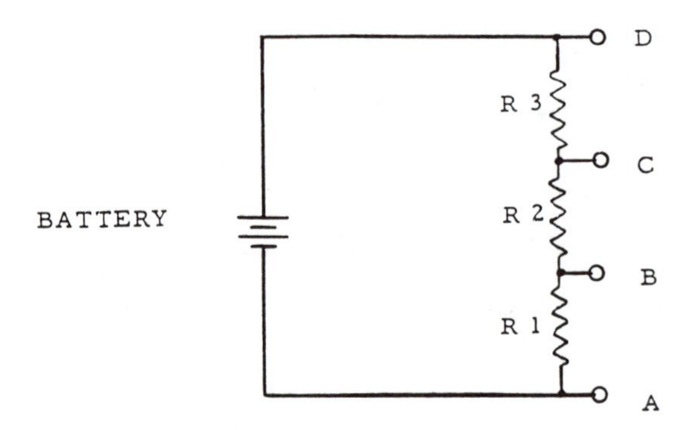

Figure 8.5. Circuit with three resistors in series.

adds up to the voltage of the source. This fact can be expressed by *Ohms law:*

$$V = IR\ , \tag{8.1}$$

where V is the voltage across the resistor (of resistance R ohms) when a current I flows through it. Applying Ohms law to the circuit in Figure 8.5, the voltage across each of the resistors is:

$$V_1 = IR_1 \tag{8.2}$$

$$V_2 = IR_2 \tag{8.3}$$

$$V_3 = IR_3 \tag{8.4}$$

and

$$V_{source} = V_1 + V_2 + V_3 = I(R_1 + R_2 + R_3)\ . \tag{8.5}$$

Voltage Divider

A voltmeter connected between points A and B of the circuit in Figure 8.5 will read the voltage across R_1 which we have called V_1 expressed by equation (8.2). When the voltmeter is connected between points A and C it will read the sum of the voltages across R_1 and R_2. The voltage is expressed by:

$$V_{A-C} = V_1 + V_2 = I(R_1 + R_2)\ . \tag{8.6}$$

The voltage between A and D will be:

$$V_{A-D} = V_1 + V_2 + V_3 = I(R_1 + R_2 + R_3) = V_{source} \tag{8.7}$$

as we have shown in equation (8.5). This circuit is called a *voltage divider* since the voltage of the source is divided between the string of resistors. More resistors can be added in series to give more selections of voltage but it should be apparent that the total voltage can never be greater than the source voltage.

A useful variation of this voltage divider is illustrated in Figure 8.6. Here the battery has been replaced by an ac source and the string of resistors has been replaced by a single resitance with a sliding contact. (The formulas in the preceding section are still valid for ac signals.) This special resistance is called a variable resistor or *potentiometer*. From the preceding example, it is easy to see that the voltage between A and B can be varied between 0 and the maximum voltage of the source. This is the principle of the *volume control* in a radio and the *hearing level control* in an audiometer. When exact proportions of the ac signal must be selected, a set of fixed resistances are arranged in a circuit similar to that of Figure 8.5. This device is called a *step attenuator* and is also used to select the measurement range in a sound level meter.

The circuit described above illustrates only the simplest form of signal level control. Information on more complex circuits, such as amplifiers and filters, may be found in any good electronics text and is beyond the scope of this course.

MICROPHONES

Dynamic Microphone

One of the most widely used types of microphone is the *dynamic microphone* shown in cross-section in Figure 8.7. Sound reaching the diaphragm causes the coil of wire to cut across the magnetic field according to the frequency and amplitude of the sound signal. The induced electrical signal voltage generated in the coil corresponds in frequency and amplitude to the acoustic signal striking the diaphragm.

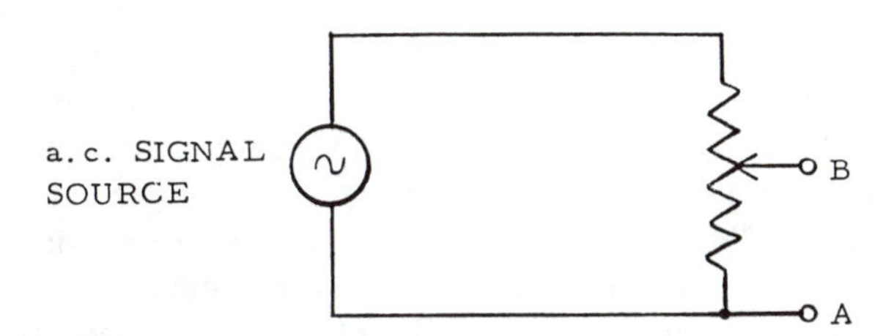

Figure 8.6. Potentiometer circuit.

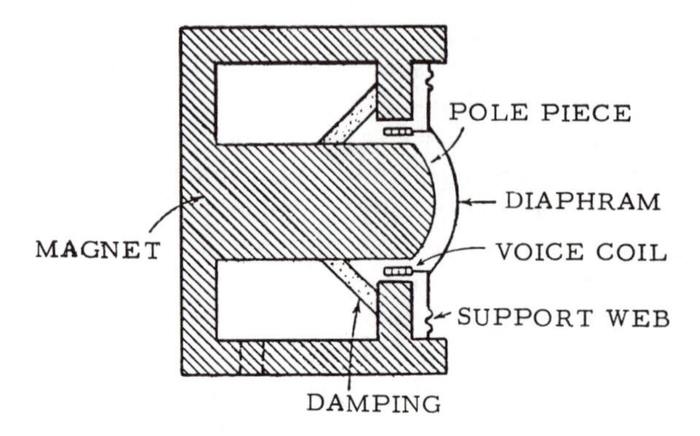

Figure 8.7. Dynamic microphone.

Thus, the acoustical energy has been transformed into an electrical signal with corresponding frequency and amplitude characteristics. The term *transducer* is applied to this device and applies in general to any device that changes one form of energy into another. The term transducer also applies to loudspeakers, which will be described later in this chapter.

Dynamic microphones can be made to have very uniform frequency response over the range of audible frequencies. They are widely used for public address systems, music recording, and radio broadcasting. Because they contain a coil of wire, however, magnetic fields from electronic equipment, such as PA amplifiers, tape recorders, and audiometers, can induce unwanted signals into this wire coil. They are not recommended for use with sound level meters where the electrical signal from the microphone is used only to activate an indicating meter. It is impossible to distinguish between this induced electrical signal and the acoustical signal when the only information available is the meter reading.

A variation of the dynamic microphone that has inherently good directional sensitivity in the *ribbon microphone* shown in Figure 8.8. Maximum electrical output is produced by sound striking the ribbon, causing it to move across the magnetic field between the pole pieces. The electrical output is minimum and approaches zero for sound arriving at the sides of the microphone so that the ribbon would tend to move parallel to the magnetic field. The resulting directional pattern resembles a figure 8. Figure 8.9 shows idealized directional patterns and illustrates the formation of a cardioid pattern from the combination of an omnidirectional pattern, such as that exhibited by an unmodified dynamic

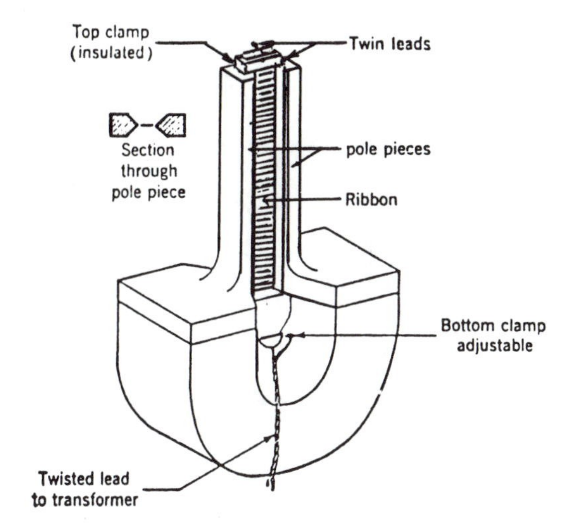

Figure 8.8. Ribbon (or velocity) microphone.

microphone, and a figure 8 pattern exhibited by a ribbon microphone. Cardioid and other directional patterns have been produced by mechanically shaping the sound inlet to an ordinary dynamic microphone element.

Piezoelectric Microphone

Certain types of crystals and ceramic materials when subjected to stress produce electrical charges on their surfaces. These charges can be collected by proper placement of conducting materials. This property is called the *piezoelectric effect.*

By mechanically coupling these piezoelectric materials to a diaphragm, a very practical and efficient transducer can be constructed (see Figure 8.10). Although the materials are quite stiff, proper coupling can produce a microphone with a sufficiently uniform frequency response over the range of audible frequencies that it is acceptable for use on sound level meters. Originally Rochelle salt crystals were used on sound level meters and indeed, specifications for the Type 2 sound level meter are based on the properties of Rochelle salt microphones. However, these crystals are easily damaged by temperatures only slightly above room temperature and are highly susceptible to damage from humidity. Many ceramic materials have been developed that also exhibit piezoelectric properties that are quite stable with wide variations in temperature and humidity. Although the electrical signal output is fairly high,

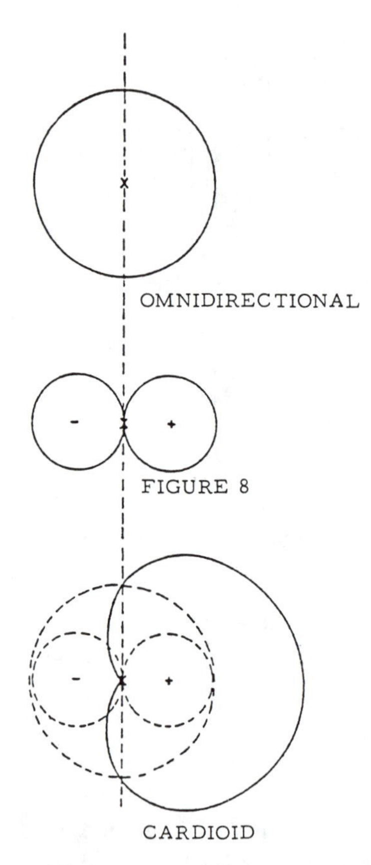

Figure 8.9. Directional patterns of microphones.

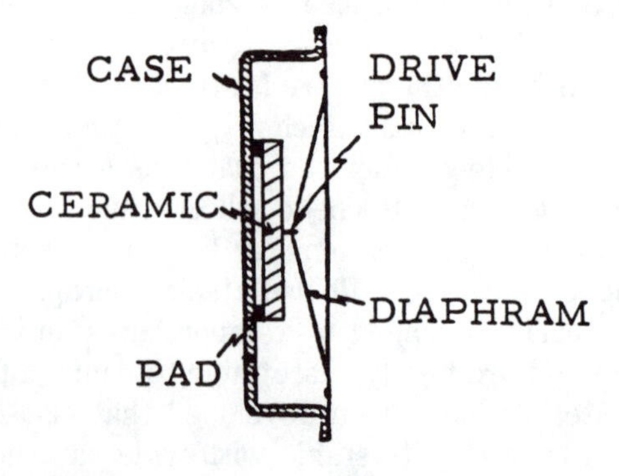

Figure 8.10. Piezoelectric microphone.

connecting cables longer than four or five feet can cause a loss of high frequency response and/or sensitivity. For operation over long cables, such as in a broadcast studio or on a concert stage, dynamic microphones are a better choice. Dynamic and piezoelectric microphones are the only microphones that transform acoustic energy directly into electrical energy without the need for a voltage or current supply to the microphone element.

Condenser Microphone

Figure 8.11 shows a sectional view of a precision condenser microphone. The term *condenser* for these microphones is still used by most

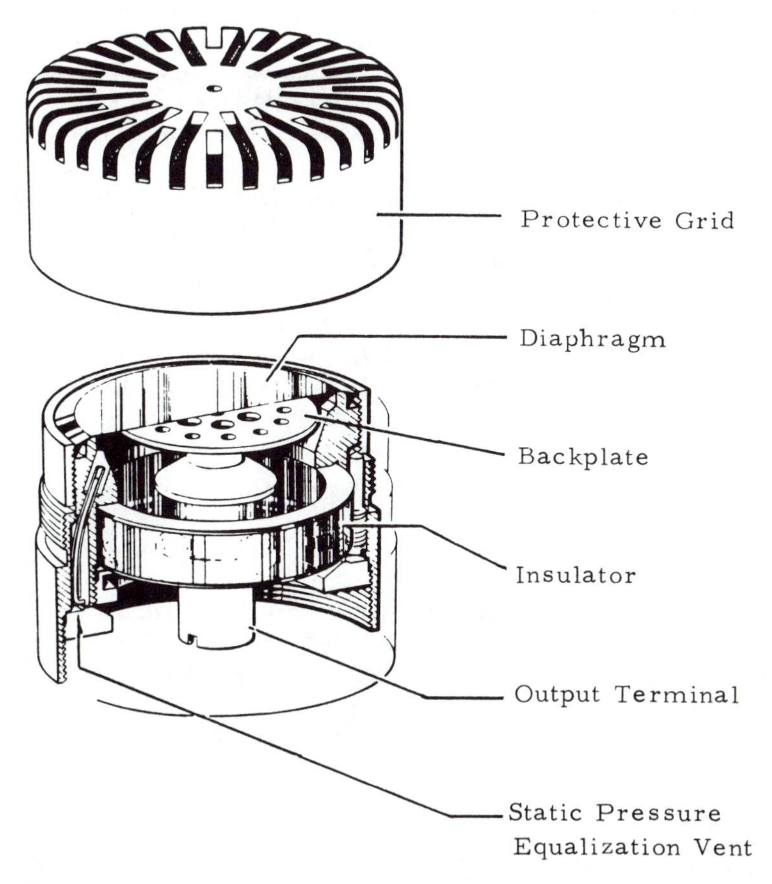

Figure 8.11. Cut-away view of a 1-inch condenser microphone.

manufacturers although technically the correct term should be *capacitor* microphone. In operation, an electrical capacitor is formed between the conducting diaphragm and the rigid, insulated back plate. A supply of electricity is connected through a very high resistance to the back plate, resulting in an electric charge on the capacitor formed by the diaphragm and the back plate. The changes in capacitance due to sound causing the diaphragm to vibrate are detected by measuring the change in charge on this capacitor. To accomplish this, a very sensitive amplifier must be located close to the microphone element. While the need for an amplifier and voltage supply close to the microphone element appears to be a disadvantage, the very good stability and excellent frequency response of this type of microphone make it the ideal choice for use on precision sound level meters. Condenser microphone elements can be built for frequencies up to 140 kHz and down to 2.5 Hz. The frequency response can be tailored to give uniform frequency response when used in a cavity-type coupler ("artificial ear") or when used in a free diffuse sound field. These types of microphones are designated *pressure type* and *free-field type* respectively. Their use is discussed in the chapter on sound level measurement.

The *electret microphone* is a variation of the condenser microphone in which the electrical charge used to supply the capacitive element is permanently installed within the diaphragm material. A sensitive preamplifier is still needed very close to the microphone element so that the signal can be connected to an amplifier through a cable. The primary advantage is that the entire package is very low in cost and the frequency response is quite uniform. Many models have been built with sufficient long-term stability so that they are acceptable for use on sound level meters. Their low-cost construction is demonstrated by their widespread use on inexpensive cassette tape recorders.

Carbon Microphone

All of the microphones mentioned above are *reciprocal transducers*. This refers to the fact that they can be used not only to change acoustic energy to electrical energy but, when suitably connected to a signal source, they can also transform electrical energy into acoustic energy. This is actually done in reciprocity calibration of microphones.

A detailed description of reciprocity calibration is given in *Fundamentals of Acoustics* by L. E. Kinsler, A. R. Frey, A. B. Coppers and J. V. Sanders, Wiley & Sons (1982). (In reality, a loudspeaker is much more suitably designed to transform electrical energy into acoustic energy, as you will see later in this chapter.)

The carbon microphone, shown in Figure 8.12, is one of the most frequently used microphones; it is an excellent example of a microphone that is *not* reciprocal. For many years, the carbon microphone has been the only one used in telephones although new electronic, lightweight telephones now use other types of microphones.

The carbon microphone used for the sensing element a small packet of carbon granules in contact with the diaphragm. Movement of the diaphragm compresses these granules to reduce the electrical resistance or relaxes pressure on the granules to increase the electrical resistance. The changes in resistance corresponds in frequency and magnitude to the acoustical signal vibrating the diaphragm. A current of electricity passing through the packet of carbon granules therefore varies according to the changes in resistance; see Ohm's law, equation (8.1). This varying electrical signal represents the highest electrical output obtainable from any microphone, although the signal-to-noise ratio obtained is the poorest of all microphones. Only in rare situations is this high electrical noise signal a detriment to the primary purpose of the telephone—that of voice communication.

The microphones described in this chapter are those most often encountered in actual use. Other more sophisticated principles have been applied to acoustic transducers but these are still experimental. Nearly all transducers in general use operate on some simple variation of these principles.

LOUDSPEAKERS

Although the operating principles of loudspeakers are similar but reciprocal to those of microphones, a major difference is apparent:

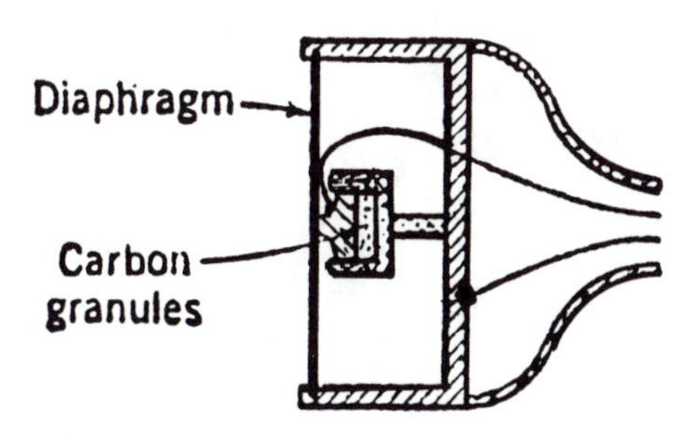

Figure 8.12. Carbon microphone.

loudspeakers are designed to transform large amounts of electrical *power* to the movement of air.

The most widely used loudspeaker, the dynamic type, is shown in Figure 8.13. Schematically, it looks very much like the dynamic microphone with some major differences: the diaphragm is made in the shape of a cone and is much larger than the diaphragm of a microphone. The cone shape allows a very low mass system to compress a large amount of air. In a loudspeaker, the voice coil is made of much heavier wire than that used in a microphone so that large amounts of electrical power can be applied.

All of the principles used for microphones have been applied to loudspeakers with various advantages and disadvantages. In a microphone the mass of the moving system is made as small as possible so that it can respond to the minute pressure variations encountered over a wide range of frequencies. Indeed, a single microphone element can easily be made to be uniformly sensitive to the entire range of audible frequencies from 20 to 20,000 Hz. While the very small microphones used for supersonic frequencies do not respond well to the lowest audible frequencies, they do respond to an extremely wide range of frequencies.

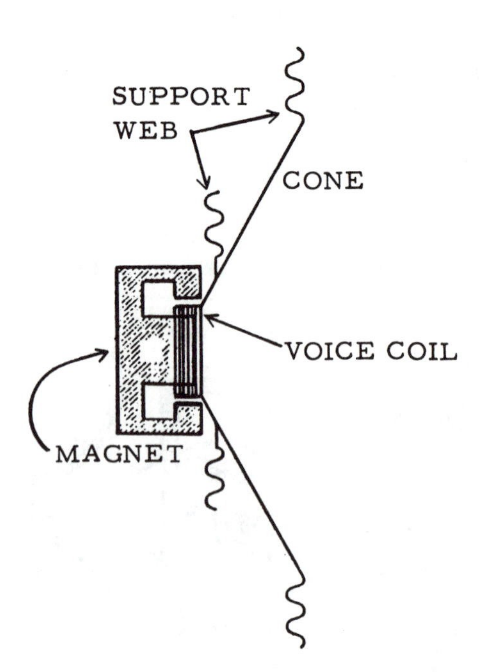

Figure 8.13. Dynamic loudspeaker.

In contrast, it is not possible to build a single loudspeaker element that will reproduce all frequencies equally well. For this reason, many different designs using different operating principles are employed to reproduce different parts of the audible spectrum. To reproduce the full range of audible frequencies from 20 to 20,000 Hz, a group of two or three different types of loudspeakers are often assembled in a single mounting together with appropriate electrical filters so that each element is used for only that range of frequencies it is most suited to reproduce. These electrical filters are called *crossover networks*. The complete system is commonly referred to as a two-way or three-way speaker system. The individual loudspeaker elements in a system are given descriptive names such as *woofer*, *midrange* and *tweeter*. Each element may operate on the same or completely different principles.

The dynamic loudspeaker described above is especially suited for reproducing low frequencies and is nearly always used as the woofer in a loudspeaker system. At middle and high frequencies, the mass of the moving element prevents efficient response so that a different design must be used to accommodate different parts of the audible spectrum. Often it is possible simply to design a smaller size unit with lower mass to favor response to higher frequencies. Indeed, many three-way systems employ three sizes of dynamic loudspeaker elements. The operating principles of the condenser microphone and piezoelectric microphone have been applied to the design of high frequency loudspeakers or tweeters.

Condenser loudspeakers are often referred to as *electrostatic* speakers. Their inherently low mass moving systems are a distinct advantage in reproducing high frequencies, although the electrical impedance of these elements is a disadvantage that can be overcome only with special circuitry. Electrostatic woofers have been built but have not proven to be commercially acceptable.

As we have noted in Chapter 4, the directionality of a source is proportional to the size of the radiating surface and inversely proportional to the wave length. Applying this principle to a loudspeaker, it is apparent that for a given loudspeaker diameter, the directionality gets larger (beam width gets smaller) as the wave length becomes smaller. In other words, the beam width becomes smaller as frequency increases. Therefore, if a woofer could reproduce high frequencies, the beam width would be so narrow that the full frequency range could be heard only in a narrow cone in line with the axis of the speaker. This is another good reason for using smaller size speakers for midrange and tweeters. This is illustrated in Figure 8.14. Here the beam width of the woofer is assumed to be about the same as that of the tweeter at the

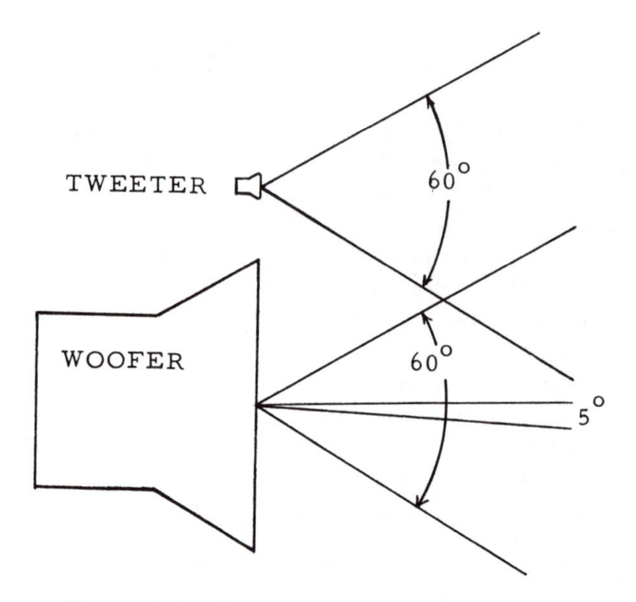

Figure 8.14. Directionality of loudspeakers.

low frequency end of their individual operating range. In the far field where the beam patterns overlap, the full acoustic spectrum can be heard. For a listener located in the near field on the axis of the woofer, the high frequencies from the tweeter will be greatly reduced in level. If the woofer were capable of reproducing the frequencies that are normally sent to the tweeter (which is not possible), the directionality

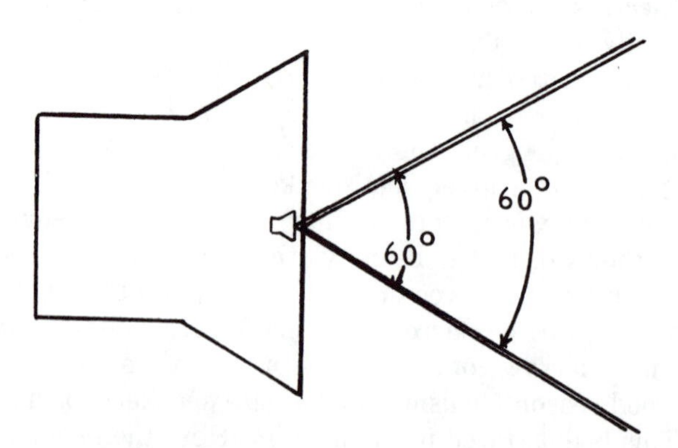

Figure 8.15. Coaxial loudspeaker system with matched beam patterns.

pattern would become extremely narrow so that a listener slightly off the axis of the woofer would hear the high frequency information at reduced level. This could have a detrimental effect on speech discrimination. Note that the proper design of woofer, midrange and tweeter can be made to give approximately equal beam widths over the individual operating ranges. In the far field (a long distance away) of a three-way loudspeaker, the three elements blend to produce a uniform field over the full frequency range. However, in the near field (close to the speaker), a listener may not hear all parts of the audible spectrum with the proper balance. Phase changes at the edges of the radiation pattern could cause serious distortion of the composite sound. This situation can occur with distributed loudspeaker systems (two- or three-way systems) when used in a small listening room such as an audiometric booth. To overcome this problem, some loudspeaker systems are built with the different elements mounted coaxially (see Figure 8.15). With this arrangement, a listener in the near field will be located on the axis of each loudspeaker element. When distributed-element loudspeakers must be used with a listener in the near field, it is advisable to locate the listener on the axis of the midrange speaker (or tweeter in the case of a two-way system) since most of the speech frequencies are reproduced by this element.

It is always desirable to mount a loudspeaker in some sort of box to improve its radiation efficiency, to provide mechanical protection and to improve its appearance. Since the radiation efficiency and low frequency response are affected by the design of the box for a particular loudspeaker, it is usually desirable to consider the box a part of the loudspeaker system.

Indeed, it is possible to design a loudspeaker box or baffle that will enhance reproduction of certain types of musical instruments such as stringed instruments or woodwinds. One of the most important characteristics for speech reproduction is the transient response of the speaker system. Since speech contains so much transient energy, any resonance of the loudspeaker system will be excited giving a *coloration* to the reproduced sound. This is especially noticeable when a simple loudspeaker is mounted in an open-back box such as are often used in interoffice paging systems. In this type of installation, the principal resonance of the speaker box becomes dominant for any voice signal being reproduced. Usually the best sound reproduction is obtained when the box or baffle is designed specifically for a given set of loudspeakers. A much more detailed discussion of loudspeakers and baffle designs may be found in engineering and "high fidelity" textbooks.

CHAPTER 9
SOUND MEASUREMENT

THE CONCEPT OF LEVEL

Sound Power and Sound Power Level

Sound power is the term used to describe the amount of acoustic energy that is produced per unit time by a sound source. It is measured in watts (W). Since the actual sound power encountered in practical situations can be as small as .000,000,001 watt or as large as 50,000,000 watts, it is convenient to use the decibel notation to express the power relative to an agreed-upon *reference power*. The sound power is then converted to the same dimensions as the reference power so that a *dimensionless ratio* can be formed. (The mathematical operation to be carried out requires this ratio to have no dimensions such as watts or newtons.) The *decibel* unit (dB) is then defined by the following relationship:

$$L_w = 10 \log \frac{W}{W_o} \qquad (9.1)$$

where L_w is the power *level* in decibels (dB), W is the sound power in watts, and W_o is the reference power in watts. The arbitrarily chosen reference power is 10^{-12} watt. Table 9.1 shows the relationship between sound power in watts and sound power level in dB relative to 10^{-12} watt and may be referred to as *the level in dB above 10^{-12} watt.*

Note that the numbers in the center column of Table 9.1 are much simpler to use than the actual sound power in the left-hand column. It can be demonstrated mathematically that no precision is lost in using this simplification. The term *level* may be compared to the floors of a tall building as observed while riding in the elevator. Each floor represents a certain level above the bottom floor which is analogous to the reference quantity. For example, the sound power of normal voice conversation can be thought of as being at a level of seventy floors above the ground floor or reference power. Similarly, the sound power of a seventy-five piece orchestra is at a level of sixty floors above that of a voice conversation. Acoustically, we say that the power level of the seventy-five

Table 9.1.
Typical Power Levels for Various Acoustic Sources.

Power (Watts)	*Power Level (dB re 10^{-12} Watt)*	*Source*
25 to 40 million	195	Saturn Rocket
100,000	170	Ram Jet
10,000	160	Turbo-Jet Engine
1,000	150	
100	140	
10	130	75-Piece Orchestra
		Pipe Organ
1	120	Small Aircraft Engine
		Large Chipping Hammer
		Piano
0.1	110	Blaring Radio
		Centrifugal Ventilating Fan
0.001	90	Auto on Highway
		Vaneaxial Ventilating Fan
		Voice—Shouting
0.0001	80	
0.00001	70	Voice—Conversational Level
0.000001	60	
0.0000001	50	
0.000,000,01	40	
0.000,000,001	30	Voice—Very Soft Whisper

piece orchestra is 60 dB above the level of conversational speech. Likewise, it can be observed that a shouting voice is at a level of 20 dB above conversational level and a whisper is about 40 dB below conversational level.

The term decibel (dB) is commonly used as the unit for level rather than the fundamental unit bel so that decimal notation is unnecessary for most applications. For example, 195 decibels is the same as 19.5 bels. However, in certain situations requiring extra-precision measurements, it is advisable and customary to express levels to one-tenth of a decibel.

Sound Intensity and Sound Intensity Level

Sound intensity at any specified location may be defined as the average acoustic energy per unit time passing through a unit area that is normal to the direction of propagation. Sound intensity is, therefore, related to sound power by:

$$I = \frac{W}{4\pi r^2} \tag{9.2}$$

where I is the average sound intensity in watts per unit area, W is the sound power in watts, and r is the distance from the source.

The quantity $4\pi r^2$ is the area of a sphere surrounding the source over which the intensity is averaged. It is obvious from equation (9.2) that the intensity will decrease with the square of the distance from the source; hence, the well-known inverse-square law.

Sound intensity units, like sound power units, cover a wide range, so again here it is desirable to use the dB notation to compress the measuring scale. The intensity level is defined as:

$$L_I = 10 \log \frac{I}{I_0} \text{ dB} , \tag{9.3}$$

where I is the measured intensity at some given distance from the source, and I_0 is a reference intensity. The reference intensity commonly used is 10^{-12} W/m^2.

Sound Pressure and Sound Pressure Level

Very complex procedures are necessary to measure sound power and sound intensity; consequently, it is easier to measure sound pressure, since the output of most microphones is directly related to the sound pressure on the microphone diaphragm. Since the pressure varies above and below the atmospheric pressure, this varying component of the pressure is expressed as the square root of the mean square of the instantaneous pressures measured over a full cycle of pressure variation. This varying component is written as the *root mean square* (rms) pressure. Whenever we see the expression *sound pressure*, the rms. value is assumed unless specifically stated otherwise.

Just as with sound power and sound intensity, the range of sound pressures is very wide. Some commonly encountered sound pressures are shown in Figure 9.1. Sound pressures near the pain threshold (about 20 N/m^2) are found in many work areas; audiologists commonly work with sound pressures near the threshold of hearing (about 0.00002 N/m^2). The sound pressure corresponding to the threshold of human hearing has been chosen as the reference pressure for calculation of sound pressure level. Other standard reference quantities are listed in Table 9.2.

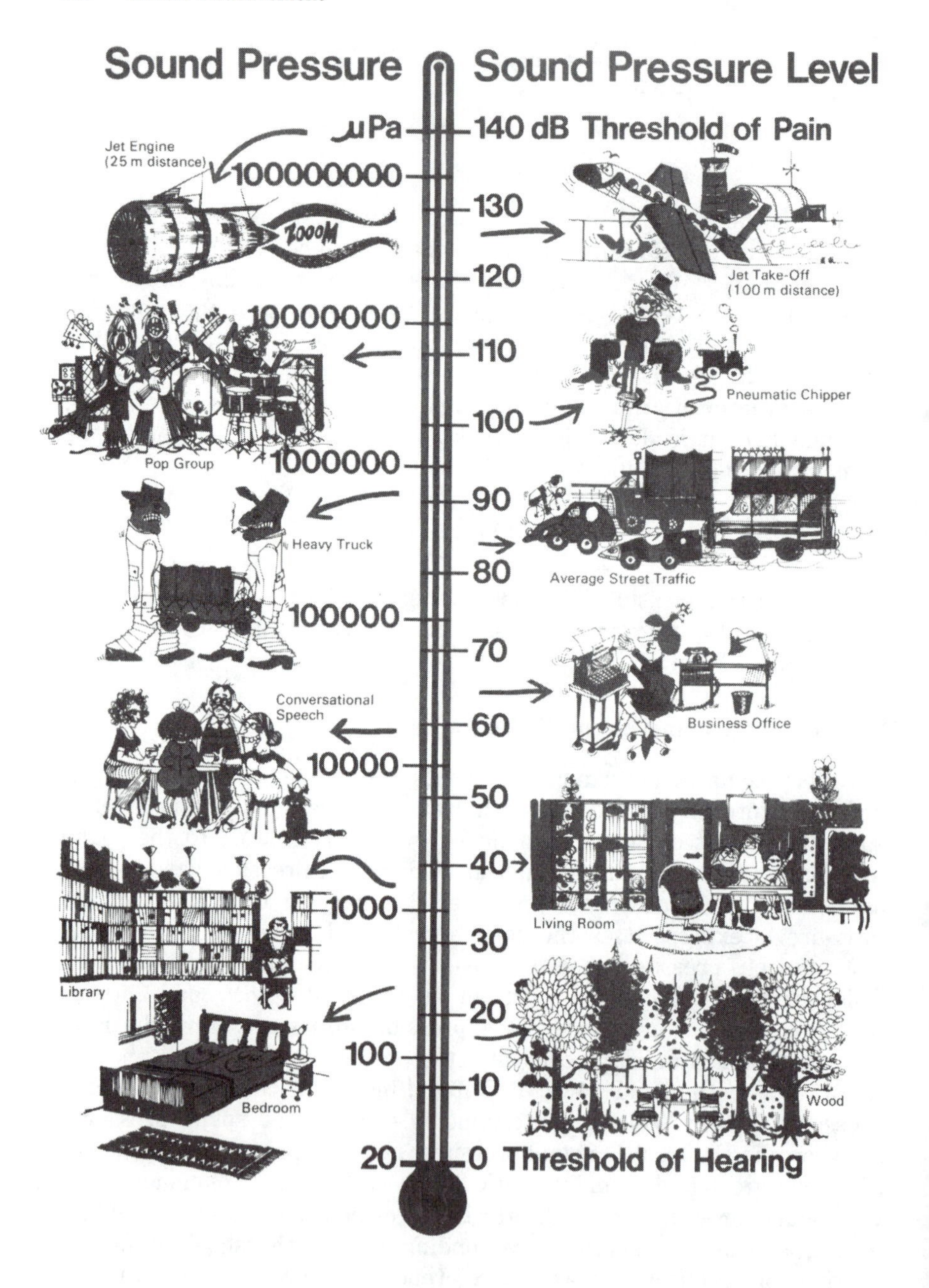
Sound Pressure
Sound Pressure Level
µPa
140 dB Threshold of Pain
Jet Engine
(25 m distance)
100000000
130
ZOOOM
Jet Take-Off
(100 m distance)
120
10000000
110
Pneumatic Chipper
100
Pop Group
1000000
90
Heavy Truck
80
Average Street Traffic
100000
70
Conversational
Speech
Business Office
60
10000
50
40
1000
Living Room
30
Library
20
100
10
Bedroom
Wood
20
0 Threshold of Hearing

Figure 9.1.

Table 9.2.
Preferred Reference Quantities for Various Acoustical Levels.

		SI Unit	*CGS Unit*
Gases:	$P_0 = 20\ \mu N/m^2$	$= 2 \times 10^{-5}\ N/m^2$	$= 2 \times 10^{-4}$ dyne/cm^2
Liquids:	$P_0 = 1\ \mu N/m^2$	$= 1 \times 10^{-6}\ N/m^2$	$= 1 \times 10^{-5}$ dyne/cm^2
	$a_0 = 10\ \mu M/s^2$	$= 1 \times 10^{-5}\ m/s^2$	$= 1 \times 10^{-3}\ cm/s^2$
	$v_0 = 10$ nm/s	$= 1 \times 10^{-8} m/s$	$= 1 \times 10^{-6}$ cm/s
	$F_0 = 1\ \mu N$	$= 1 \times 10^{-6}$ N	$= 1 \times 10^{-1}$ dyne
All Media:	$P_0 = 1$ pW	$= 1 \times 10^{-12}$ W	$= 1 \times 10^{-5}$ erg/s
	$I_0 = 1\ pW/m^2$	$= 1 \times 10^{-12}\ W/m^2$	$= 1 \times 10^{-9}$ erg/(s · cm^2)
	$E_0 = 1\ pJ/m^3$	$= 1 \times 10^{-12}\ J/m^3$	$= 1 \times 10^{-11}$ erg/cm^3
	$W_0 = 1$ pJ	$= 1 \times 10^{-12}$ J	$= 1 \times 10^{-5}$ erg

By definition, the decibel is a dimensionless unit: the logarithm of the ratio of a measured quantity to a reference quantity. The decibel is commonly used to describe levels of acoustic intensity, acoustic power, hearing thresholds, electric voltage, electric current, electric power, etc., as well as sound pressure levels; thus, it has no meaning unless a specific reference quantity is specified.

Most sound-measuring instruments are calibrated to provide a reading of root mean square (rms) sound pressures on a logarithmic scale in decibels. The reading taken from such an instrument is called the *sound pressure level* (L_p).

For sound measurements in air, 0.00002 N/m^2 is commonly used as the reference pressure. Often the reference pressure is stated in pascals and since one pascal (Pa) is defined to be equal to one newton per square meter, the reference pressure may be stated as 20 micropascals (20 μPa).

The formula for sound *pressure* level is slightly different from the formulas for sound *power* level and sound *intensity* level. Recall from Chapter 4, equation (4.4), that the sound power is related to the sound pressure by the formula:

$$\frac{W}{4\pi r^2} = \frac{p^2}{\rho c}\,. \tag{9.4}$$

Accordingly, the sound power responding to the reference acoustic pressure (pref) can be written using equation (9.4) as:

$$\frac{W_{\text{ref}}}{4\pi r^2} = \frac{(P_{\text{ref}})^2}{\rho c}\,. \tag{9.5}$$

Solving equations (9.4) and (9.5) for the power gives:

$$W = \frac{4\pi r^2}{\rho c} p^2 \tag{9.6}$$

and

$$W_{\text{ref}} = \frac{4\pi r^2}{\rho c} (p_{\text{ref}})^2 . \tag{9.7}$$

Substituting these expressions in equation (9.1) gives the following expression for power level:

$$L_W = 10 \log \frac{W}{W_{\text{ref}}} = 10 \log \frac{\frac{4\pi r^2}{\rho c} p^2}{\frac{4\pi r^2}{\rho c} (p_{\text{ref}})^2} . \tag{9.8}$$

Since equations (9.4) and (9.5) refer to conditions at the same point in the same medium, equation (9.8) can be simplified to:

$$L_W = 10 \log \frac{p^2}{(p_{\text{ref}})^2} = 10 \log \left(\frac{p}{p_{\text{ref}}}\right)^2 . \tag{9.9}$$

Mathematically this can also be written as:

$$L_W = 20 \log \frac{p}{p_{\text{ref}}} . \tag{9.10}$$

Since this is actually an expression for power level in terms of sound pressure, we give it the designation *sound pressure level* and write it as:

$$L_p = 20 \log \frac{p}{p_{\text{ref}}} \tag{9.11}$$

where p is the measured rms sound pressure, p_{ref} is the reference sound pressure, and the logarithm (log) is to the base 10. Thus, L_p should be written in terms of decibels referenced to a specified rms sound pressure. For example, in air the notation for L_p is commonly abbreviated as dB re 20 μPa. This reference pressure of 20 μPa corresponds closely with the reference power of 10^{-12} watt.

Figure 9.1 shows the relationship between sound pressure (in micropascals) and sound pressure level (in dB re 20 μPa) and illustrates the advantage of using the dB scale rather than the wide range of direct pressure measurements. It is of interest to note that any pressure range over which the pressure is doubled is equivalent to 6 dB whether at high or low levels. For example, a range of 20 to 40 μPa, which might be found in hearing measurements, and a range of 1,000,000 to 2,000,000 μPa, which might be encountered in community noise studies, are both ranges of 6 decibels.

The L_p referenced to 20 μPa may be written in any of the following six forms:

$$\begin{aligned} L_p &= 20 \log (p/0.00002) \\ &= 20 (\log p - \log 0.00002) \\ &= 20 \{\log p - (\log 2 - \log 10^5)\} \\ &= 20 \{\log p - (0.3 - 5)\} \\ &= 20 (\log p + 4.7) \\ &= 20 \log p + 94 \text{ dB re } 20\ \mu\text{Pa}. \end{aligned}$$

RELATIONSHIP OF POWER, INTENSITY, AND PRESSURE

Many noise control problems require a practical knowledge of the relationship between pressure, intensity, and power. An example would be the prediction of sound pressure levels that would be produced around a proposed machine location from the sound power data provided for the machine.

For a spherical or free-progressive sound wave, the intensity may be expressed by

$$I = \frac{p^2}{\rho c} \tag{9.12}$$

where p is the rms sound pressure, ρ is the density of the medium, and c is the speed of sound in the medium. The product ρc is called the characteristic impedance of the medium and is often expressed in rayls, a unit named after the famous acoustician Lord Rayleigh. It is obvious from this definition that sound intensity describes, in part, characteristics of the sound in the medium, but does not directly describe the sound itself.

Example: The manufacturer of a pneumatic chipping hammer states that the hammer has an acoustic power output of 1.0 watt. Predict the sound pressure level (L_p) at a position 30 meters from the hammer.

Assuming free-field conditions and an omnidirectional source, with the help of equations (9.2) and (9.12), the power output (W) can be expressed by:

$$W = I_{avg} \times 4\pi r^2 = \frac{p^2_{avg} \times 4\pi r^2}{\rho c} \tag{9.13}$$

where ρc for air at 20°C = 415 rayls. Solving for p_{avg}:

$$p_{avg} = \sqrt{\frac{W\rho c}{4\pi r^2}} \,. \tag{9.14}$$

Substituting the values from the example, we get:

$$\begin{aligned} p_{avg} &= \sqrt{\frac{1.0 \text{ watt} \times 415 \text{ rayls}}{4\pi \times (30 \text{ meters})^2}} \\ &= \sqrt{3.67 \times 10^{-2}} \\ &= 1.92 \times 10^{-1} \text{ N/m}^2 \,. \end{aligned}$$

Then,

$$\begin{aligned} L_p &= 20 \log \frac{0.192 \text{ N/m}^2}{0.00002 \text{ N/m}^2} \\ &= 79.6 \text{ dB re } 20\ \mu\text{Pa} \qquad (0.00002 \text{ N/m}^2 = 20\ \mu\text{Pa}) \end{aligned}$$

The actual noise level may be higher when many echoes are present.

SOUND LEVEL METERS

The sound level meter (SLM) (see Figure 9.2) is the basic instrument for measuring the overall sound pressure level of continuous or moderately fluctuating kinds of sounds. A sound level meter consists of a microphone, calibrated amplifier-attenuator circuits, frequency weighting networks, and an indicating meter. The microphone transforms the acoustic signal received at its diaphragm to an equivalent electrical signal with the same frequency and amplitude characteristics. The weighting networks modify the frequency spectrum of the electrical signal with selective characteristics patterned after those of the human ear. This *frequency weighting* therefore provides the means whereby the measured level of the sound may be correlated to the perceived level. The carefully calibrated amplifier-attenuator circuits provide a

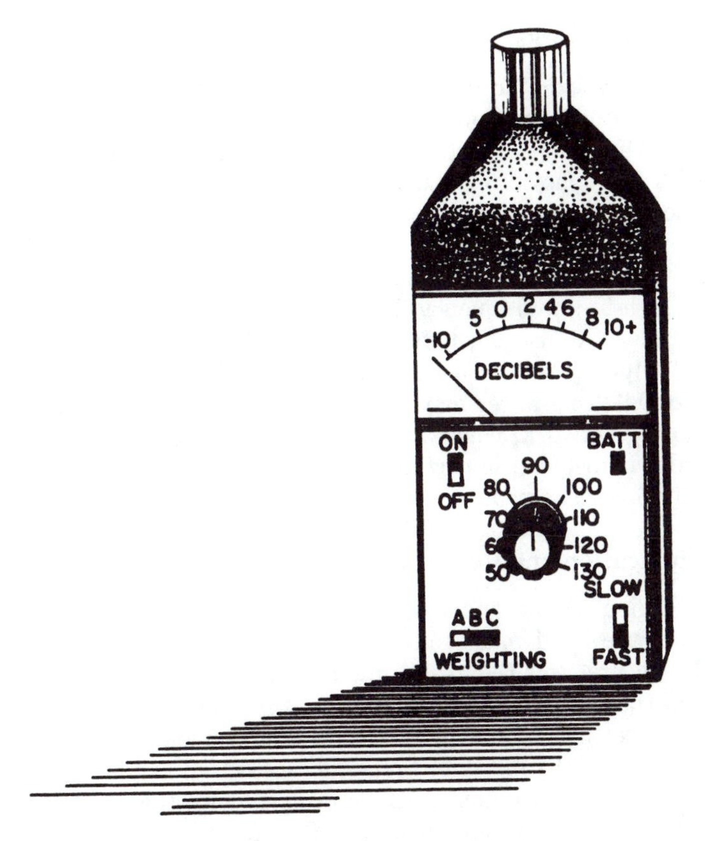

Figure 9.2. Basic sound level meter.

signal to the indicating meter where the sound level is displayed in decibels.

The operational characteristics of a sound level meter are specified by both national and international standards. The American National Standards Institute, ANSI S1.4-1983 "Specification for Sound Level Meters" (1) provides the maximum allowable tolerances used for most applications in the United States, and for the various types of sound level meters (such as Precision Type 1 and General Purpose Type 2).

Several special-purpose sound level meters are available which read directly in energy equivalent levels, L_{eq}, with averaging times ranging from 1/8 second to 24 hours. Some have a standby or pause control to stop integration for a given time period.

Weighting Networks

Sound level meter frequency weighting networks were originally intended to provide reasonable correlations between meter readings and loudness. They can also determine roughly how sound energy is distributed with frequency. In community noise measurements and industrial noise measurements, the most often used *A-weighting* gives good correlation with human response. Differences between the A-weighted and C-weighted (or flat-weighted) levels afford a good approximation of the ratio of high-to-low frequency distribution of the sound.

The ideal A-, B- and C-frequency weightings, relative to a flat or overall frequency response, as specified by ANSI S1.4-1983, are shown in Figure 9.3. Tolerances may be found in the ANSI Specifications. The D-weighting network, which emphasizes frequencies between 1,000 and 10,000 Hz, is included in certain special sound level meters and is used primarily for noise measurements around airports.

Meter Indication and Response

According to the above ANSI standard, the indicating meter or readout of the SLM must have a scale covering a range of at least 15 dB. The accuracy of the scale graduations must be at least ± 0.2 dB except in the lower part of the scale that is overlapped by a change in attenuator setting where the accuracy requirement is ± 0.5 dB. The response time of the indicator (generally measured as the response time of the complete SLM) must be in accordance with the FAST or SLOW dynamic characteristics specified. The FAST response specifications for a Type 1 instrument require the meter to indicate within 0 to 4 dB less than the reading for a continuous signal of 1,000 Hz with a duration of 200 milliseconds. The SLOW response specifications require the meter to be within 2 to 6 dB less than the continuous signal reading for a Type 2 instrument and 3 to 5 dB less than the continuous signal reading for a Type 1 instrument for a 1,000 Hz signal with a duration of 500 milliseconds.

If sound level fluctuations are rapid but of a duration of 500 milliseconds or longer, the SLM may be used with reliable accuracy. With the exception of impulsive sounds, most noises may be measured with the FAST or SLOW meter characteristics. FAST meter characteristics should be used wherever possible for the greatest accuracy; however, when the sound levels are fluctuating rapidly, the SLOW meter characteristics may be used to get reproducible readings. The SLOW response averages the sound input so that there are smaller ranges of level change and the rates of change are reduced so that the meter can be

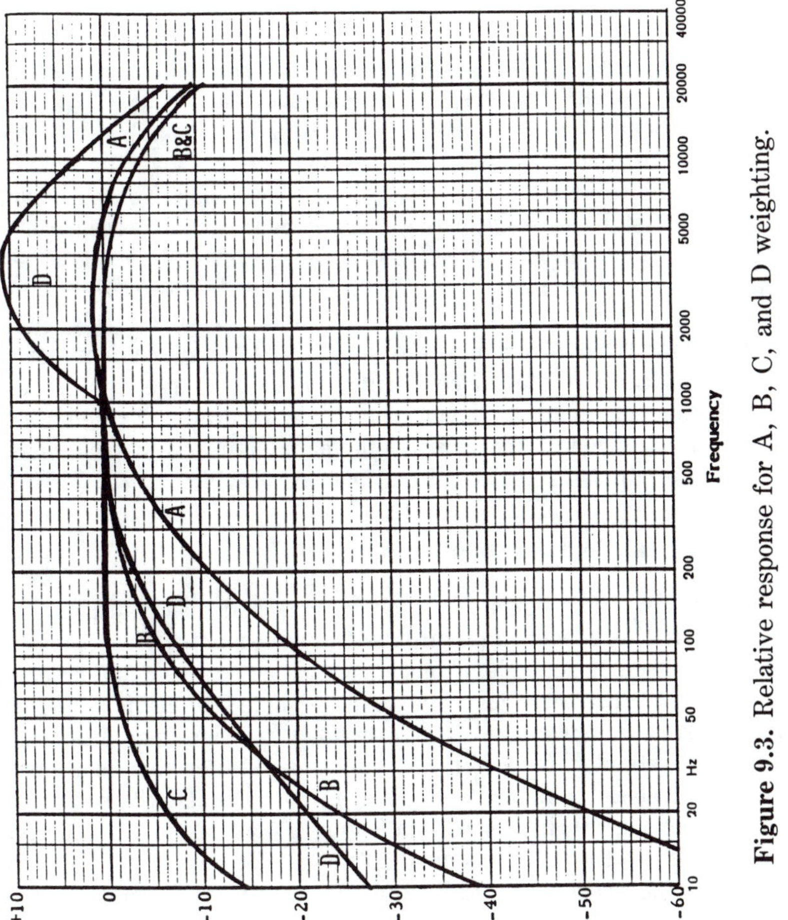

Figure 9.3. Relative response for A, B, C, and D weighting.

read more accurately. The SLOW response is particularly useful when widely fluctuating sound levels are to be compared from one time to another (i.e., before and after noise control measures). If the sound level is fluctuating 6 dB or less, a subjective judgment of central tendency ("eyeball average") is usually acceptable. If the sound level is fluctuating more than 6 dB, manual or automatic sampling of sound levels may be required.

When impulsive sounds such as those from gun shots, pile drivers, drop forges, or jack hammers, are encountered, an oscilloscope or an impulsive-type SLM must be used (2,3,4). Impulsive sounds are considered to be those whose sound pressure levels rise above the ambient by 10 dB or more in a time less than 200 milliseconds. The measuring instrument must be capable of reading the peak sound pressure level (unweighted). If an impulse-type SLM is used, it should include a peak indicating and holding circuit so that the peak level is held long enough to be read or until manually reset. Peak sound pressure levels should be recorded for at least 10 impulses in close succession so that a numerical average level can be determined. Generally, the average, the highest, and the range of impulsive levels should be recorded. Extreme care must be taken to follow the instrument manufacturer's instructions so that accurate impulse sound level data can be obtained.

Microphones for Sound Measurement

Each type of microphone has advantages and disadvantages that depend upon the specific measurement requirements. Two types of microphones used on sound level meters are called *free-field* and *pressure* microphones. A pressure microphone is designed to have a uniform frequency response when used in a coupler (such as the Bureau of Standards 9A coupler) for calibrating earphones. When this microphone is used in a free field, as for sound level measurements, the sensitivity increases above 2,000 Hz as shown in Figure 9.4 when the sound arrives perpendicular to the microphone diaphragm. However, as shown in Figure 9.4, the response is quite uniform to 10,000 Hz if the sound arrives in the plane of the microphone diaphragm (grazing incidence). Pressure type microphones are the best choice for sound arriving at random incidence. A free-field microphone, on the other hand, is designed to have the most uniform response for sound arriving perpendicular to the diaphragm (0°). They are especially useful for measurements in anechoic chambers where plane waves are arriving from a controlled direction and there are no reflecting surfaces present.

Calibration and frequency-response curves and stability character-

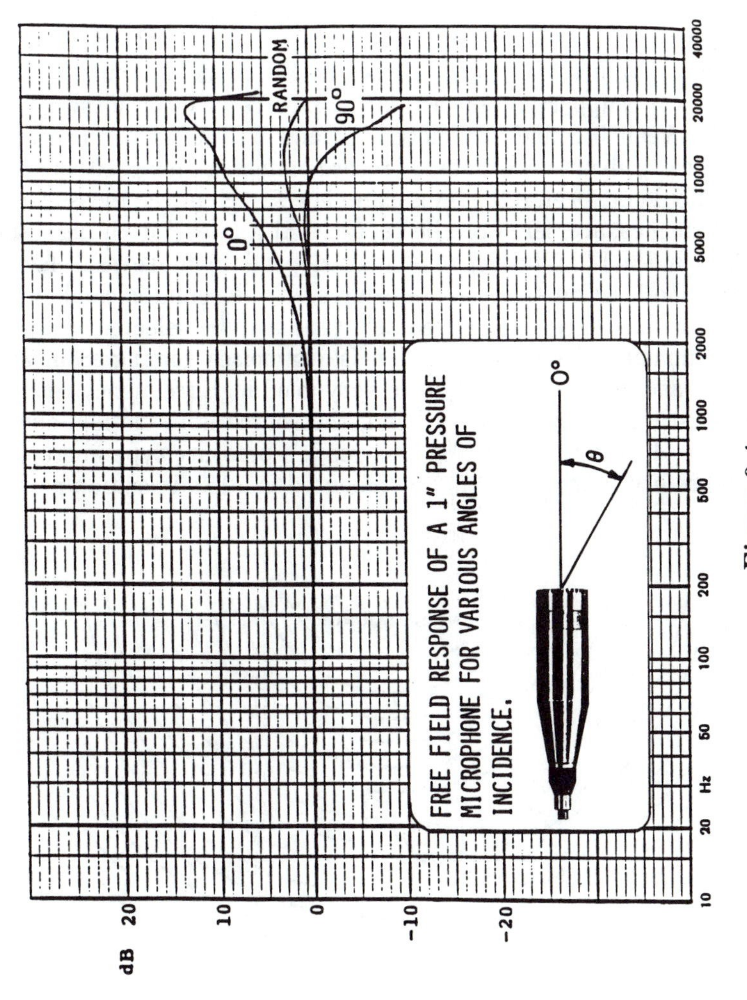
FREE FIELD RESPONSE OF A 1" PRESSURE MICROPHONE FOR VARIOUS ANGLES OF INCIDENCE.
0°
θ
RANDOM
90°
dB
20
10
0
−10
−20
10
20
Hz
50
100
200
500
1000
2000
5000
10000
20000
40000

Figure 9.4.

istics with respect to temperature, humidity, vibration and electromagnetic fields are generally available from the instrument manufacturer. Performance limitations for the microphone system may be found in ANSI S1.4-1983 for Types 1 and 2 sound level meters.

Any microphone must be oriented as specified in the manufacturer's instructions; otherwise errors will result that will be particularly prominent at high frequencies. The most frequently specified orientation is illustrated in Figure 9.5. The preferred height of the microphone above the ground or supporting surface is 1.2 meters (4 feet), although any height between 0.6 and 1.8 meters (2 and 6 feet) is acceptable for specific measurement conditions. A record of microphone position should be carefully kept, preferably on a plan view sketch of the measurement site so that measurements can be repeated at a later date if necessary.

The choice of a microphone may depend upon several factors, including the location of the sound source. If the sound is coming from a particular fixed direction, a free-field type may be selected because it will discriminate against potential masking noises coming from other directions and generally it will have very good high frequency response characteristics. If, on the other hand, the source is in motion, such as in the case of a vehicle traveling on a road, a pressure type microphone may be preferred because it can be mounted in a fixed position pointing

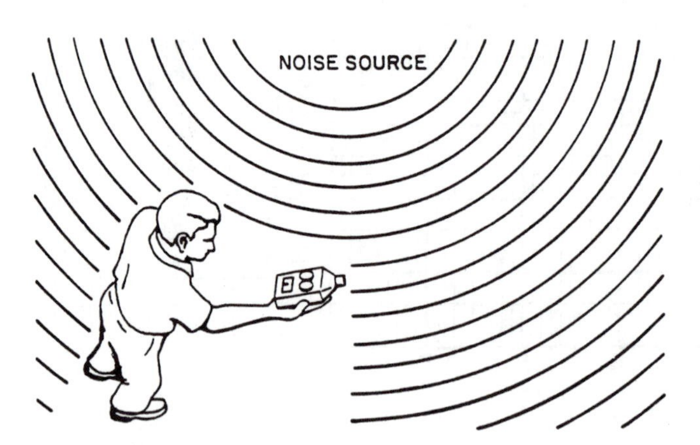

Figure 9.5. The sound level meter should be oriented with respect to the source of sound as recommended in the SLM instruction manual for the microphone being used. Many microphones should be pointed at a right angle to the sound path as shown here.

upward and receive the sound at grazing incidence as the vehicle moves. The microphone calibrated for random incidence is generally a good choice for measurements in a diffuse sound field where the sound is coming from all directions. These microphones may be used interchangeably in most situations, but the manufacturer's instructions regarding orientation must be followed in each situation or errors will result.

Temperature and Humidity

Most modern microphones are not permanently damaged by normal ranges of temperature and humidity. However, temporary erroneous readings may result from condensation if the microphones are moved from very cold to very warm areas. To avoid errors from condensation, the instruments should be turned on and allowed to sit in the measurement area for at least five minutes prior to making measurements. Temperature and humidity correction curves are generally supplied with the microphone and should be consulted.

Microphone Cables

In most noise assessment situations, sound level measurements should be made with the microphone mounted on the sound level meter. However, there are special situations where an operator's body, or even the instrument case, should be removed from the measurement area to obtain accurate data.

When extension cables are required for microphones, care must be taken to make the necessary corrections to the sound level reading according to the instrument manufacturer's instructions. Some microphones require special electronic circuitry when used with cables and others do not. The amount of correction for given lengths of cable also varies from one instrument system to another. Therefore, the manufacturer's instructions should be followed precisely. Finally, the microphone must also be calibrated while it is mounted on the cable before and after it is used.

Windscreen

Rapid air movement over a microphone causes turbulence that in turn generates extraneous noise. This noise can effectively mask the sound being measured and cause erroneous high level readings. The use of earphones connected to the SLM output jack (consult manufacturer's recommendations) often will enable the operator to detect wind-generated noise; however, low level masking may occur that will be inaudible.

It is good practice to use microphone windscreens in any case when wind or wind gusts are suspected during the course of measurements.

Generally, windscreens are either spherical or cylindrical shaped open-celled polyurethane, or silk-covered grids attached directly over the microphone. However, there are limits to their effectiveness. Three "rules of thumb" are:

1. Measurements should never be made, even with windscreens, in winds having velocities greater than 20 km/hr (12 mph).
2. Measurements should not be made if wind noise is audible through a monitoring headset connected to an SLM when using the A-weighting and the lowest attenuator setting (setting for measuring the lowest sound level to be measured).
3. Measurements may be made using a windscreen and a frequency-band analyzer as long as it can be determined that the wind noise remains at least 10 decibels below the sound being measured in any of the frequency bands.

In all cases, the windscreen should be one provided by the SLM manufacturer for that instrument. Corrections should be available for these windscreens. If such a windscreen is not available, if no corrections are available for a windscreen, or if a windscreen is old or soiled, tests should be made by presenting reproducible sounds to the microphone with and without the windscreen in place. The test sound used should contain low, medium, and high frequency components (i.e., 500, 1,000, 2,000, 4,000 and 8,000 Hz). If possible, the windscreen should be tested with similar frequency components to those expected from the sounds to be measured. Corrections should be developed and used for differences up to 2 dB. If the windscreen causes changes greater than 2 dB, the windscreen should be discarded.

Calibration

Two kinds of instrument calibration procedures must be used to ensure that accurate measurements are obtained. A laboratory calibration should be performed at regular intervals not more than one year apart. These calibrations should be done by the instrument manufacturer or qualified personnel at acoustical laboratories. Equally important field calibrations should be made before and after each use of the measurement equipment. Field calibrations are conducted with acoustic calibrators provided by the instrument manufacturers.

Generally, the field calibrators are compact, battery operated devices that provide a means for conducting an overall system calibration check. Some calibrators generate a single frequency and others provide several different test signals, all at specified sound pressure levels. Field calibrators are designed to be used on specific microphones and they should be used only on these microphones. Otherwise, error may result or microphones may be permanently damaged.

In use, the sound level generated by the calibrator should correspond to the SLM reading. If it does not, consult the instrument instruction book to determine how adjustments should be made. Calibrations are usually made using the flat- or C-weighting settings on the SLM unless otherwise specified by the manufacturer.

Caution should be exercised when using calibrators at atmospheric pressures different from that at sea level. Normally, correction data are supplied by the instrument manufacturers.

COMBINING SOUND LEVELS

It is often necessary to calculate the overall sound level from a group of sound level measurements. For example, the noise from a random-noise source is often measured in sets of overlapping *frequency bands* to determine the distribution of energy radiated at different frequencies. The resulting measurement is called an *acoustic spectrum* and is analogous to the rainbow produced when light is passed through a prism, hence the term "spectrum". When the center frequencies (c.f.) of these bands are in multiples of two, they are called *octave bands*. The edges of these frequency bands overlap just so that all frequencies present are measured. The overall sound pressure level (spl) can be calculated by combining the levels measured in each of these bands. The procedure is useful in estimating the total sound pressure level resulting when a machine with a known noise spectrum is placed in a room with other machines already present whose spectra are known.

Since sound pressure levels are logarithmic quantities, simple addition of levels in dB constitutes a multiplication of the pressure ratios, a meaningless result. The sound pressure level resulting from a combination of sources must be calculated from the total sound power or the total sound intensity of the combination.

Generally, noise contains a wide distribution of frequencies with nearly random phase relationships. Sound pressure levels of random noises can be combined by converting the levels to intensity units that may then be added arithmetically. The combined intensity can then be reconverted to sound pressure levels in dB. Equation (9.3) can be used in free-field conditions for this purpose.

Any number of sound pressure levels can be combined, providing that all are calculated using the same reference quantity. For simplicity, let us calculate the combined sound pressure level resulting from two sound sources operating at the same time. At a given observation point, the sound pressure levels can be expressed as an intensity ratio using equation (9.3):

$$L_1 = 10 \log \frac{I_1}{I_{ref}} \tag{9.15}$$

and

$$L_2 = 10 \log \frac{I_2}{I_{ref}} . \tag{9.16}$$

Note that I_{ref} is used in the denominator instead of I_0. Although there is a calculable difference between the reference level for sound pressure and the reference level for sound intensity, it is necessary only to express the sound levels in intensity *units* and since these intensity units are all expressed in ratio form with a common denominator, the ratios can be easily combined with no loss of accuracy. Solving equations 9.15 and 9.16 for the intensity ratios gives:

$$\frac{I_1}{I_{ref}} = 10^{(L_1/10)} \tag{9.17}$$

and

$$\frac{I_2}{I_{ref}} = 10^{(L_2/10)} . \tag{9.18}$$

The intensity ratios on the left can be combined to find the intensity ratio corresponding to both sources operating at the same time:

$$\frac{I_1 + I_2}{I_{ref}} = 10^{(L_1/10)} + 10^{(L_2/10)} . \tag{9.19}$$

The combined sound pressure level for the two sources is then:

$$L_p = 10 \log \frac{I_1 + I_2}{I_{ref}} . \tag{9.20}$$

Although this is a somewhat simplified calculation, it does give the correct result. Note that the sound pressure level L_p has been taken to be equivalent to the intensity level L_I and that the equations all involve ten times the logarithm of an intensity ratio. Although the intensity reference may not be exactly equivalent to the pressure reference, the *same* intensity reference is used to obtain the final answer so that the result is exactly the sound pressure level for the combined sound. Although only two sources were combined in this example, any number of sources can be combined by adding the calculated intensity ratios, assuming that all calculations were made with the same reference value.

Although the above method gives an accurate calculation of the combined level of multiple sources, it is a somewhat cumbersome procedure. When a thorough understanding of the principles of adding sound energies has been developed, the procedure can be simplified by the use of a table similar to the one shown in Table 9.3. The error encountered in using this table seldom exceeds 1 dB, so it is usable for most situations, and the limited number of values makes it relatively easy to commit to memory. The use of this table may be illustrated by the following example:

> The sound pressure level produced by a random noise source is measured in eight discrete octave bands. The measurements are 85, 88, 91, 94, 100, 97, 90 and 88 dB respectively. These octave band levels can be combined two at a time to obtain a single value for the overall sound pressure level. It is best to begin with the lowest levels since the combined lower levels may add significantly to the total. To use Table 9.3, levels must be combined *two at a time*. Individual values must be used *only once*. In our example, then, the levels 85 and 88 dB differ by 3 dB; therefore, opposite the range 2 to 4 in the left-hand column of Table 9.3 read a value of 2 in the right-hand column. This value is then *added* to the *higher* of the two levels: 88 + 2 = 90 dB. This result is now added to the

Table 9.3.
Simplified Table for Combining Decibel Levels of Noise with Random Frequency Characteristics.

Numerical Differences Between Levels	*Amount to be Added to the Higher Level*
0 – 1	3
2 – 4	2
5 – 9	1
> 10	0

next level by repeating the process: 91 − 90 = 1; therefore, from the table we find that 3 dB must be added to the highest level. Thus, 91 + 3 = 94 dB. This procedure is continued with each octave band level to arrive at an overall sound pressure level of 104 dB.

The overall sound pressure level calculated in the above example corresponds to the value that would be found by reading a sound level meter at this location with the frequency weighting set so that each frequency in the spectrum is weighted equally. Common names given to this frequency weighting are: *flat*, *linear*, *20 kHz* and *overall*.

Many noise regulations require that the sound level be expressed as the *A-weighted* sound pressure level. The correct notation for this is L_A although the notation dB(A) is popularly used. This A-weighting can be applied to the octave band levels in the above example before combining the band levels to calculate the overall A-weighted level. To avoid errors in calculation, a systematic approach is recommended as follows:

First make a table of the levels as follows:

Octave Band Center Frequency (Hz)	31.5	63	125	250	500	1,000	2,000	4,000
Sound Pressure Levels (dB)	85	88	91	94	95	100	97	90

Now make the adjustment for A-weighting:

A-Weighting (dB)	−39	−26	−16	−9	−3	+0	+1	+1

These values are then numerically added to the octave band levels to obtain A-weighted band levels:

A-Weighted Band Levels	46	62	75	85	92	100	98	91

These A-weighted octave band levels may now be combined using Table 9.3 as described above to obtain the A-weighted sound pressure level which is about 103 dB(A). Another example of this type of calculation is given in Chapter 11.

Most industrial noises have random frequency characteristics and they may be combined as described in the above paragraphs. However, there are a few cases of noises with pitched or major pure-tone components where these calculations will not hold, and phase relationships must be considered. In areas where pitched noises are present, standing waves will often be recognized by rapidly varying sound-pressure levels over short distances. It is not practical to try to predict levels in areas where standing waves are present.

REFERENCES

1. *American National Standards Institute Specification for Sound Level Meters*, ANSI S1.4-1983.
2. Peterson, A. P. G. and E. E. Gross, Jr., *Handbook of Noise Measurement*, General Radio Co., Concord, MA, Seventh Edition, form No. 5301-8111-K, 1982.
3. Broch, J. T., *Acoustic Noise Measurements*, Bruel & Kjaer, 1975.
4. Sataloff, J. and Michael, P. L., *Hearing Conservation*, Charles C. Thomas, 1973.

CHAPTER 10
ARCHITECTURAL ACOUSTICS

For most people, the term *architectural acoustics* brings to mind the design and performance of concert halls, auditoriums, and churches. Indeed, this aspect of the field is well covered in popular magazines and technical journals, but the acoustical design of living and working spaces and testing environments is equally important and challenging. Proper acoustical design of classrooms is essential for the most effective teaching at all levels. However, in spite of the popularity of the subject, it is always the hope of architectural acousticians that persons occupying or using the spaces are not aware of the acoustics.

In audiology, we are usually concerned with living and working spaces having relatively small volumes such as testing rooms and classrooms. The two most important acoustical aspects of these small rooms are privacy and speech intelligiblity.

PRIVACY

To prevent sound in one area from intruding or interfering with activities in another area, the acoustical transmission properties of the walls, floor and ceiling must be taken into consideration. Depending upon the activities being carried out, extraneous sounds entering the space must be so reduced in level that they do not distract and, in the case of audiology, cannot be heard.

This last aspect of privacy implies that any sound entering the working space such as a test room must be below audibility. Since the threshold for speech is about 10 to 20 dB sound pressure level (spl) and since normal conversation is in the range of 60 to 70 dB spl, it is easy to calculate that the sound passing through the walls into the test room must be reduced by 40 to 50 dB.

Tables of acoustical transmission loss for architectural materials, such as Table 10.1, show that this reduction could be accomplished by an 8 inch thick brick wall. It should be noted that Table 10.1 shows

Table 10.1.
Characteristics of Certain Architectural Materials.
Data is for 1/3-Octave Band at 500 Hz.

Material	*Transmission Loss(dB)*	*Transmission Coefficient* (τ)	*Absorption Coefficient* (a)
Walls:			
Wood paneling on stud	30	.001	.17
Plasterboard on stud	38	.0003	.05
Lath and plaster	35	.0003	.04
Concrete block painted	45	.00003	.06
Brick [4 inches thick]	45	.00003	.02
Brick [8 inches thick]	50	.00001	.02
Floor and Ceiling Material			
Acoustical tile	10	.1	.53
Wood floor/plaster ceiling	35	.0003	.09
Concrete slab	40	.0001	.01
Asphalt tile on concrete	40	.0001	.03
Carpet on concrete	40	.0001	.50
Windows:			
Single glazed	20	.01	.18
Double glazed [0.5-inch air]	35	.0003	.18
Double glazed [1.0-inch air]	45	.00003	.18
Double glazed [6.0-inch air]	55	.000003	.18
Doors:			
Hollow core	15	.03	.15
Solid core	20	.01	.10
Weather stripped solid core	25	.003	.10

data for a 1/3-octave band of noise centered at 500 Hz. The acoustical properties of architectural materials are actually a function of frequency as the example in Fig. 10.1 shows. However, acoustical performance of various structures can often be compared simply by referring to the properties at 500 Hz. It is apparent from studying the table that this amount of sound attenuation of 40 to 50 dB is not easily accomplished and requires careful acoustical design.

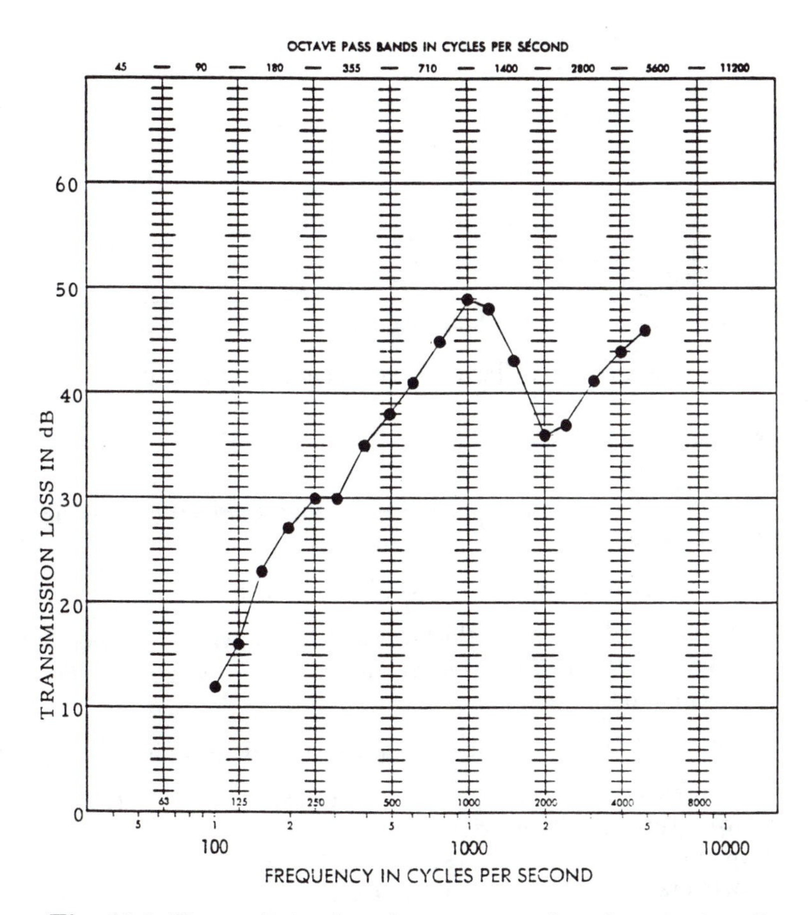

Fig. 10.1. Transmission loss for a gypsum board-on-stud wall.

SOUND TRANSMISSION THROUGH A WALL

Assume that we want to calculate the acoustic transmission loss of a wall separating two rooms as in Figure 10.2. For simplicity, assume also that the only path for sound to pass from one room to the other is through the wall in question. Assume also that the sound field is diffuse—that is, that the sound pressure is uniform over the entire wall. In Chapter 4, we learned that when sound strikes a barrier, part of the

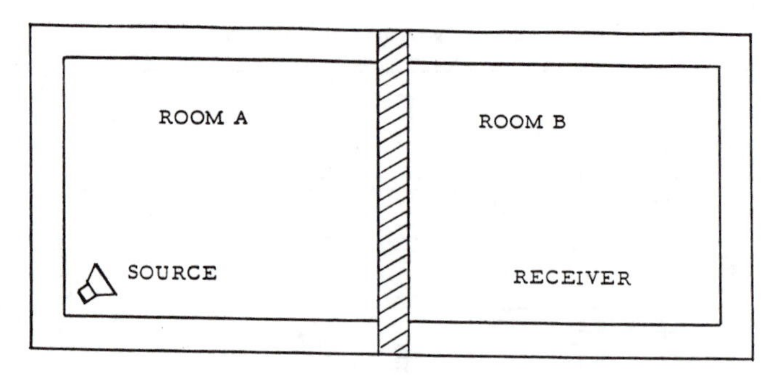

Figure 10.2. Two rooms separated by a wall.

energy is reflected, part is transmitted through the barrier according to the simplified equation:

$$I \text{ incident} = I \text{ reflected} + I \text{ transmitted} \tag{10.1}$$

When the barrier has a finite thickness, such as a wall, a fraction of the incident energy is absorbed (converted to heat energy) so that the complete equation becomes:

$$I \text{ incident} = I \text{ reflected} + I \text{ absorbed} + I \text{ transmitted}. \tag{10.2}$$

For our present purpose, we are concerned only with the energy transmitted through the barrier.

Now, in a real situation, assume that the wall also includes a door and a window according to Figure 10.3. Since the sound energy is uniformly distributed over the wall, a fraction of it will pass through the door material, another fraction will pass through the window materials, and the remaining fraction will pass through the wall material. These fractions can be calculated from the area ratios as follows:

First, calculate the total area of the composite wall:

$S = 10 \times 20 = 200 \text{ ft}^2 .$

The area of the door is:

$S_D = 3 \times 7 = 21 \text{ ft}^2 .$

The area of the window is:

$S_G = 2 \times 4 = 8 \text{ ft}^2 .$

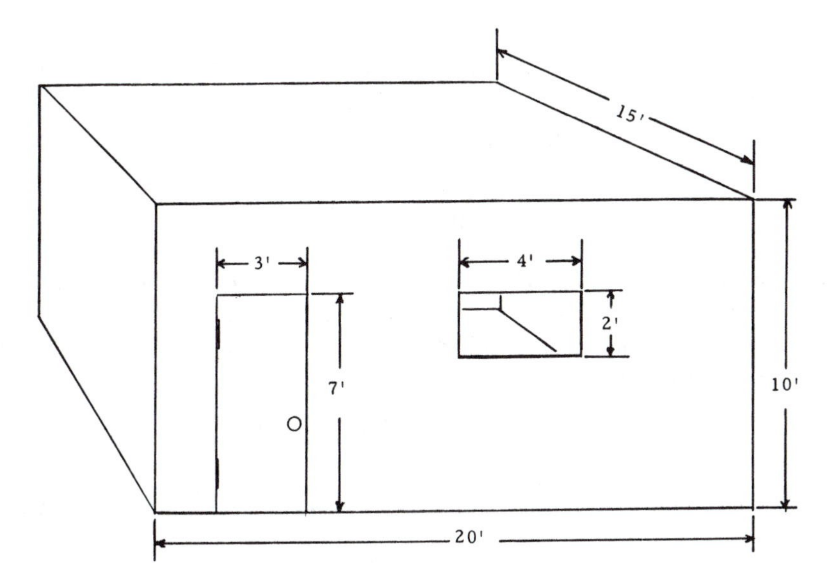

Figure 10.3. Room with door and window.

Since each of the materials will have their own peculiar transmission properties for sound, the actual area of wall material must then be:

$$S_W = S - S_D - S_G$$
$$S_W = 200 \text{ ft}^2 - 21 \text{ ft}^2 - 8 \text{ ft}^2$$
$$S_W = 171 \text{ ft}^2 .$$

Now it is necessary to know the acoustical transmission properties of the material used in this wall so again refer to Table 10.1. For this example, consider a brick wall 8 inches thick with a hollow core door (as found in most apartments) and a single glazed window. The transmission coefficient of the complete wall structure will be made up of the proportional tranmission coefficients of the individual parts as follows:

$$\tau = \frac{S_W}{S}\,\tau_W + \frac{S_D}{S}\,\tau_D + \frac{S_G}{S}\,\tau_G . \tag{10.3}$$

This equation can be written with a common denominator as:

$$\tau = \frac{S_W\,\tau_W + S_D\,\tau_D + S_G\tau_G}{S} . \tag{10.4}$$

Substituting the values from Table 10.1 and the areas we have just calculated into equation (10.4), the composite wall transmission coefficient is:

$$\tau = \frac{171 \times 10^{-5} + 21 \times 3 \times 10^{-2} + 8 \times 10^{-2}}{200}$$

$$= .003559 \text{ (or } 3.559 \times 10^{-3}) .$$

This represents the fraction of the incident energy that is actually passing through the wall. It is desirable to express this in decibels so that we can compare the actual transmission loss with the loss required in a given situation. Since:

$$\tau = \frac{I \text{ transmitted}}{I \text{ incident}} ,$$

τ is a dimensionless quantity and can be expressed in decibels as follows:

$$L_t = 10 \log \tau ,$$

where L_t is the level of the transmitted energy with respect to the incident energy. The fact that τ is always less than one results in L_t being negative.

However, since it is often convenient to express transmission loss in positive numbers, we can define:

$$TL = 10 \log \frac{1}{\tau} \tag{10.5}$$

and for the composite $T = .003559$, the transmission loss is:

$$TL = 10 \log \frac{1}{.003559}$$

$$= 24.49 \text{ dB} .$$

It is seldom necessary in a practical situation to calculate decibels to this degree of precision. However, for the purpose of demonstrating the effect of a change in architectural design, it is often desirable to retain two decimal places until arriving at the final figure. The *TL* for this example should then be reported as 24 dB since our ability to perceive differences smaller than 1 dB is very limited.

A few paragraphs back, it appeared that we should expect 40 to 50 dB transmission loss with a brick wall eight inches thick. However, in the example above, with a window and door in this wall, the transmission loss was only 24 dB. It is apparent that a considerable amount of sound energy is passing through the window and/or door. Our examination of Table 10.1 shows that some alternate window and door treatments could be chosen to yield a transmission loss closer to 40 dB or 50 dB. After choosing more appropriate window and door construction, other sound leakage paths become dominant. In a practical situation even the opening under the door and/or the holes through the wall for plumbing and electrical outlets can become the dominant leaks for acoustical energy.

Ventilation ducts must also be examined carefully for possible leakage paths for sound energy. Attention to these small leakage paths for sound energy is very important in a room designed for audiometric testing.

INTELLIGIBILITY

Within a room, sound reflections play an important role in determining intelligibility for speech. In an enclosed space where many reflections can occur with little absorption at each reflection, sound can continue to bounce back and forth with little loss in amplitude. Waves from each reflection will be delayed by their individual transmission path with the result that the original wave front can become blurred or can sound like many wave fronts each arriving at a separate time. The result is an extension of duration of the original sound such that important details of speech can no longer be distinguished or are masked causing a loss in intelligibility for speech. This phenomenon is called reverberation. An important measure of reverberation is the time required for the sound to decay through a level change of 60 dB. This measurement is called *reverberation time.*

Measurements of reverberation time can seldom be made for a 60 dB decay of sound level so the values are extrapolated from measurements made over a smaller range of decay.

Reverberation time can often be calculated from the volume, absorption coefficients, and surface area of the reflecting surfaces present in a room. The formula is the Sabine equation originally developed experimentally in 1922 by Wallace C. W. Sabine at Harvard University:

$$T_{60} = \frac{.05V}{A}, \qquad (10.6)$$

where V is the volume of the enclosed space, T_{60} is the time in seconds for a 60 dB decay of sound level, and A is the effective absorption of all surfaces in the space. (Equation 10.6 is valid for customary measurement units. For metric units, the expression

$$T_{60} = \frac{.161\,V}{A} \qquad (10.6a)$$

should be used, where A is expressed in metric sabins.)

The effective absorption A can be determined by the formula:

$$A = \Sigma(S_1\,a_1 + S_2\,a_2 + S_3\,a_3 + \cdots S_n a_n) \qquad (10.7)$$

where S_n is the exposed surface area and a_n is the absorption coefficient of surface material n.

A partial listing of absorption coefficients for some architectural materials is included in Table 10.1.

To illustrate the use of equations (10.6) and (10.7), what would the reverberation time be if Room B in Figure 10.1 was 15 feet deep with acoustical tile on the ceiling and a concrete slab floor? Use the same brick wall with the same window and door. We have already calculated the areas of wall, window and door:

$$\begin{aligned} \text{Wall } S_1 &= 171 \text{ ft}_2 \quad a_1 = .02 \\ \text{Door } S_2 &= 21 \text{ ft}^2 \quad a_2 = .15 \\ \text{Window } S_3 &= 8 \text{ ft}^2 \quad a_3 = .18\,. \end{aligned}$$

The values of (a) were taken from Table 10.1.

Assume that the other three walls are plasterboard-on-stud construction.

$$\begin{aligned} S_4 &= 10 \times 15 \times 2 + 10 \times 20 \\ &= 500 \text{ ft}^2 \qquad a_4 = .05\,. \end{aligned}$$

For the acoustical ceiling,

$$S^5 = 20 \times 15 = 300 \text{ ft}^2 \quad a_5 = .53$$

and for the concrete floor

$$S_6 = 20 \times 15 = 300 \text{ ft}^2 \quad a_6 = .01\,.$$

Using the values in equation (10.7), the total absorption for the room is:

$$\begin{aligned} A &= (171 \times .02) + (21 \times .15) + (8 \times .18) \\ &\quad + (500 \times .05) + (300 \times .53) + (300 \times .01) \\ A &= 3.42 + 3.15 + 1.44 + 25 + 159 + 3.0 \\ A &= 195.01\,. \end{aligned}$$

The reverberation time in seconds is then calculated from equation (10.6) as follows:

$$T_{60} = \frac{.05 \times 3{,}000 \text{ ft}^3}{195.01}$$

$$T_{60} = .77 \text{ seconds} .$$

According to the criteria in Table 10.2, this room would be acceptable for speech. Of course, the reverberation time would be further reduced by the presence of people, furniture and draperies, all of which should be included in the calculation of equation (10.7) to obtain a more accurate estimate of the actual reverberation time.

NOISE REDUCTION

The principles used to determine transmission loss through walls and reverberation time within rooms illustrate some very basic principles of sound reduction of a noisy source. These principles apply to any noise source whether it is part of a factory production line or a machine used in an office setting. The goal in any case is to reduce the level of the sound reaching the listeners' ears.

First, refer to Table 10.1 and note that the transmission loss in dB increases as the mass of the wall material increases. Under floor and ceiling material, note that all the treatments involving a concrete slab have about the same transmission loss as the heavier wall materials. Now, examine carefully the transmission loss data for the lightweight materials such as the doors and especially the acoustical tile ceiling. It should be apparent that *mass* is required to block the transmission of sound. The greater the mass, the more transmission loss. It should be obvious that lightweight materials, such as acoustical tile or fiberglass blankets, are ineffective as sound *barriers*.

But aren't fiberglass blankets and acoustical tile good materials for

Table 10.2.
Acceptable Reverberation Time.

Use	T_{60}
Speech	0.5 to 1.0
Music	1.5 to 2.0
Pipe Organ	2.0 to 3.0

noise reduction? The answer to this is "yes"—when they are used the proper way. To understand the proper use of these materials, refer to the column of Table 10.1 showing the absorption coefficients. These figures represent the fraction of incident sound energy that is absorbed by the material in the process of reflection. Under conditions of sound reflection, the table shows that sound is more readily *absorbed* by reflection from lightweight, fluffy materials such as acoustical tile, carpeting or wood paneling. The use of this information will become apparent by recalling the calculation of reverberation earlier in this chapter. Equation (10.6) shows that the reverberation time is inversely proportional to the total amount of absorption present within the enclosed space. The sound energy being reflected back and forth within the enclosure can combine with the sound energy being generated by the source. The resulting noise level inside the enclosure can be 3 dB to 6 dB greater than the noise generated by the source in free space—the noise inside the space is actually louder than it would be if no enclosure were present. This means that we actually have more sound energy to be attenuated by the barrier material.

These principles of noise reduction can be summarized by a simple example. Figure 10.4 shows a noisy machine mounted on a hard floor. The sound level meter represents the ear of a listener in the area. As a first step, Figure 10.5 shows a rigid enclosure (barrier) built around the machine resulting in a reduction of the noise outside. However, the noise reduction is probably not as much as expected because the machine causes the floor to vibrate which, in turn, re-radiates sound outside the enclosure. This *structure-borne sound* can be reduced by mounting the machine on a resilient pad (vibration isolators) as shown in Figure 10.6.

Reverberation of sound inside the enclosure has increased the noise level as explained previously, so that the noise reduction is still not as great as expected from the transmission loss data for the enclosure material. At this point we can make use of absorbing material, such as fiberglass or acoustical tile, lining the inside of the enclosure to reduce reverberation with its associated 3 to 6 dB buildup of sound (see Figure 10.7). The absorbing material tends to produce an anechoic environment inside the enclosure for most of the higher frequency sound energy. This illustrates, in a simplified way, the fundamental principles of noise reduction:

1. *Transmission* of sound can be reduced best by the use of massive, rigid materials.
2. *Reflection* of sound can be reduced best by soft, lightweight materials.

Figure 10.4. A machine producing a high level of noise over a wide range of frequencies.

Figure 10.5. A complete rigid-walled enclosure results in a measurable reduction in the noise level.

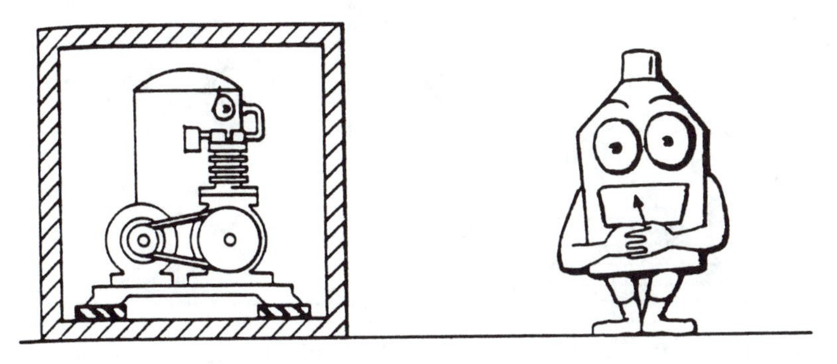

Figure 10.6. Vibration isolators under the machine inside the enclosure help to reduce structure-borne sound.

3. *Structure-borne sound* can be reduced by resilient mounting. (Detailed use of resilient mounting for vibration isolation is a complex subject and will not be covered in this text.)

For very noisy machines, it is possible to achieve some additional reduction of noise by using a double enclosure as illustrated in Figure 10.8. However, the loss of sound energy may not be twice that of a

Figure 10.7. Sound absorbing material on the inside walls reduces the buildup of sound levels within the enclosure, thereby reducing the noise levels received.

Figure 10.8. Additional noise reduction for extremely noisy sources can be accomplished by using a double-walled sound-treated enclosure.

single enclosure. In a practical situation, a considerable amount of engineering design is needed to produce a workable enclosure. Attention must be given to ventilation and power for the machine as well as arrangements to pass material in and out of the enclosure. Just as in the case of the room in the earlier example, *any* opening in the enclosure will greatly reduce the effectiveness of the sound barrier.

CHAPTER 11
ENVIRONMENTAL NOISE AND CONSERVATION OF HEARING

Noise-induced hearing loss is the most widely recognized and one of the most significant physiological effects of noise on people. It is now well established that individuals who are exposed to excessively noisy environments without adequate hearing protection measures will incur permanent and irreversible loss of hearing due to the noise exposure. However, many people do not understand the link between noise exposure and hearing loss. Individuals regularly expose themselves to high level noise and needlessly damage their hearing when the use of protective or preventive measures could have easily avoided this. This apparent lack of concern is attributable, to a great extent, to the insidious nature of noise-related hearing loss. The onset of this type of hearing loss is often very gradual, occurring over a period of years, and frequently is not noticed until the loss of hearing is considerable. Further, the symptoms of noise-induced hearing loss, such as loss of auditory sensitivity and ringing in the ears, are often deceptive. These symptoms usually subside after the period of exposure, giving the misleading impression that no permanent damage has occurred.

HAZARDOUS PROPERTIES OF NOISE

From prior research on the auditory effects of noise it is possible to list those characteristics of noise that contribute most directly to hearing loss. These characteristics are: overall noise level, frequency spectrum, exposure duration, and temporal pattern (1). Where possible, all of these factors should be considered when determining the hazard posed by a particular noise. Reliance should not be placed on a single characteristic of the noise. Differences in individual susceptibility must also be considered.

Overall Noise Level

Extended exposure to A-weighted sound levels of 70 to 80 dB will cause hearing impairment in only a few very susceptible individuals (2). It should be recognized that any specification of allowable exposure levels is a compromise based on assumptions concerning what percent of the population may realistically be protected, and concerning just what constitutes a significant hearing loss.

Frequency Spectrum

Research indicates that the ear is most sensitive to frequencies above 1,000 Hz and that hearing losses occur more readily at these higher frequencies. However, noise containing a large percentage of its energy below 4,000 Hz is considered to be more hazardous to hearing than noise containing most of its energy above 4,000 Hz (3).

Exposure Duration

Generally, as the length of exposure increases, so does the extent of the resultant hearing loss. Studies have suggested, however, that noise-induced hearing loss usually develops most rapidly during the first ten to fifteen years of exposure (4).

Temporal Pattern

The relationship between intermittent noise and hearing loss is not clearly defined. In general, however, intermittent noise has been shown to be less damaging than continuous noise, for the same total energy content. For example, four hours of continuous exposure to an L_A of 100 dB can be expected to be more hazardous than an exposure to the same sound energy one hour on and one hour off over an eight-hour day.

Summary

In summary then, the following general statements can be made concerning the hazardous properties of noise:

- the louder the noise, the more damaging it will be to hearing;
- the frequency components of noise between 1,000 and 4,000 Hz are more damaging than the low frequency components;
- generally, as the length of noise exposure increases, so does the extent of the resultant hearing loss; and

Figure 11.1 Comparison of hazardous and non-hazardous properties of noise.

- continuous noise is generally more damaging than intermittent noise for the same energy content.

HOW NOISE DAMAGES HEARING

Observations in animals as well as in people show that noise reaching the inner ear directly affects the hair cells of the hearing organ (organ of Corti). These hair cells serve an important transducing function in audition. They convert the mechanical energy reaching the ear into neuroelectrical signals, which are carried by the auditory nerve to the brain. The outer ear, eardrum, and middle ear are almost never damaged by exposure to intense noise, although in some extreme situations, the eardrum can be ruptured by very intense impulsive noises. Blasts or other very loud impulse noises can also damage the organ of Corti by causing vibrations that simply tear apart some or all of the structure. Injuries resulting from single exposures to large pressure changes are called *acoustic trauma*. As the intensity of the noise and time for which the ear is exposed are increased, a greater proportion of the hair cells and their supporting structures are damaged or eventually destroyed.

Hearing acuity is generally affected first in the frequency range from 2,000 to 6,000 Hz, with most affected persons showing a loss or "dip" at 4,000 Hz. If high level exposures are continued, the loss of hearing will further increase around 4,000 Hz and spread to lower frequencies. There is a great deal of individual variation in susceptibility to noise damage, so there is no single level of noise that separates safe and unsafe conditions for all ears. Furthermore, neither the subjective loudness of a noise, nor the extent to which the noise causes annoyance or interference with human activity, are reliable indicators of its potential danger to the hearing mechanism (5).

Indications of Noise-Induced Hearing Loss

Two noticeable indications of noise-induced damage to the auditory system are usually evident immediately following exposure to high level noise. They are:

1. ringing in the ears (tinnitus)
2. a loss of auditory sensitivity

A loss in auditory sensitivity can be determined by measuring the change in the absolute hearing threshold level. The *absolute hearing threshold level* is that at which a tone can just be detected. In other words, it represents the lower limit of our range of audibility. The greater the hearing threshold level, then, the greater the extent of hearing loss. An increase in the threshold level that results from noise exposure is called a noise-induced threshold shift. These threshold shifts

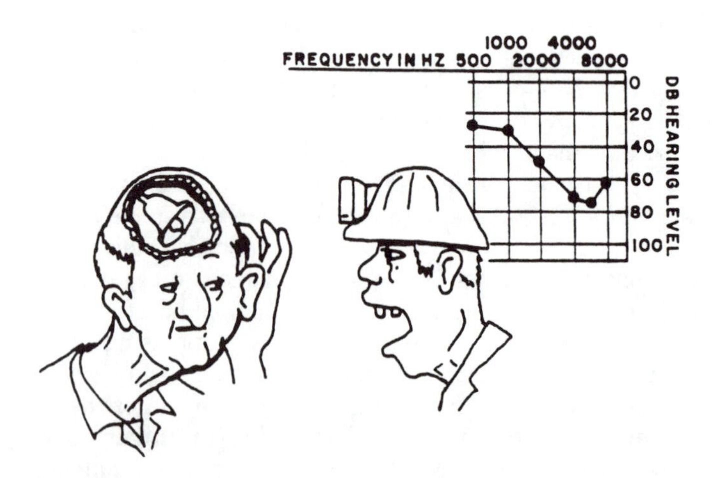

Figure 11.2 "Indications of noise-induced hearing loss."

can be either temporary or permanent. *Temporary threshold shifts* (TTS) decrease over a period of time until they disappear. *Permanent threshold shifts* (PTS) reflect changes in hearing which do not recover with time. As exposures are repeated, the ear may become less able to recover from the temporary threshold shifts and permanent hearing changes can be observed.

DETERMINATION OF A HEARING HANDICAP

The principal criterion of the extent to which hearing loss is a handicap is the ability to understand speech in quiet surroundings. However, much debate exists concerning the implications and significance of small amounts of hearing loss, and most guidelines for the assessment of the extent of handicap are based only on thresholds for tones in the region most important for the reception of speech (500, 1,000 and 2,000 Hz).

The Committee on Hearing of the American Academy of Ophthalmology and Otolaryngology (AAOO) has adopted guidelines stating that a handicap exists when the average hearing threshold level for 500, 1,000 and 2,000 Hz exceeds 25 dB in the better ear (6). However, research shows that individuals with hearing losses above 2,000 Hz may experience considerable difficulty in understanding speech in moderate levels of background noise (7), even though they do not come close to meeting the AAOO criterion. Hearing losses above 2,000 Hz impair hearing so that it is difficult to distinguish the sounds of consonants that contain much of the information required to discriminate speech sounds. Because of this, several states have now included thresholds for 3,000 Hz in the determination of significant hearing loss for their compensation laws.

PRESBYCUSIS AND OTHER FACTORS AFFECTING HEARING

Presbycusis is the term given to hearing loss specifically ascribed to the effects of aging. Hearing becomes less sensitive with advancing age, even in the absence of damaging noise exposure. This effect is most pronounced at frequencies above 3,000 Hz (8). At least in Western cultures, presbycusis appears to be more pronounced in males than in females, but this may be due to the noisy and more stressful activities that are more commonly engaged in by males.

The probability that a person will develop a hearing impairment due to noise depends on the pattern of exposure from all noises. It may be possible to control occupationally related noise exposure, but the control of non-work exposures poses a much more difficult problem. Recreational and other non-work exposures have been categorized as *sociocusis* factors (9). These factors complicate attempts to control the acoustic environment and make it very difficult to determine the long-term noise dose (over several years) that must be known in order to establish an accurate relationship between noise exposure and hearing loss.

HEARING CONSERVATION

Hearing conservation programs are designed to protect individuals from the hazardous effects of noise. Most hearing conservation programs are based on conditions at the work place; however, it is not unreasonable to extend these principles and practices to the community, where damage to hearing also occurs.

In all cases, it should be kept in mind that the objective of a hearing conservation program is to prevent noise-induced hearing loss. Simple compliance with local, state, or Federal rules and regulations generally will not prevent all noise-induced hearing loss in susceptible individuals, because the exposure limits selected for compliance purposes have by necessity been developed with consideration of the economic impact of control measures. Obviously, the lowest and safest economically feasible limits are desirable for the well-being of the individual.

An effective program should include three areas of concentration: hearing assessment, noise dose assessment and noise reduction.

Hearing Assessment

One of the most important phases of a hearing conservation program is the measurement of hearing levels of persons exposed to noisy environments. A program of periodic audiometric evaluations should be implemented and carried out by a trained audiologist. Although there are numerous audiometric tests, most hearing conservation programs rely on a pure tone absolute threshold test as their principal index of hearing sensitivity. If the audiogram indicates that losses or changes in hearing have taken place since the base audiogram was taken, then the person should be referred for professional evaluation of the change.

Assessment of Noise Dose

Noise hazard areas generally are identified by the duration and level of sound exposures. Noise dose, which depends on the product of these two factors, can either be measured using a sound level meter and a clock, or be measured directly with a personal noise dosimeter. The resultant noise doses should be at least as low as those specified by the OSHA (Occupational Safety and Health Administration, discussed later in this chapter), but should be as low as is feasible for the particular noise exposure location.

Noise Reduction

If the noise assessment indicates that hazardous conditions exist, several protective steps should be taken immediately. These include: wearing hearing protection, modifications to the source, and modifications to the path of sound transmission.

Source Modification: Attempts at source modifications usually begin with locating the source of the noise. Once located, the source should

Figure 11.3 Examples of possible modifications for protection of hearing.

be eliminated or modified to reduce the radiated sound. A detailed examination of engineering control procedures is beyond the scope of this course but many detailed presentations of this topic may be found in books on noise control (10–13). Be aware that the use of engineering control procedures on noisy equipment already in operation may be difficult and, in many cases, ineffective. Engineering noise control measures can be used most effectively at the design stage of potentially noisy equipment. Until recently there has not been a strong demand for quiet equipment, and available technology has not been used to full advantage in product design.

Path Modification: If it is not possible to obtain enough reduction of noise level by treatment of the source, the next step is to reduce the exposure level by modification of the sound path. A number of steps can be taken to reduce the production and propagation of noise (10–13). These include the use of:

1. Partial and complete barriers placed between the observer and the source to reduce the level of sound propagated (these include personal hearing protection and sound isolation enclosures).
2. Absorption materials placed on room surfaces and inside of enclosures to prevent reflection and buildup of noise levels.
3. Damping materials placed on vibrating surfaces to reduce vibration and in turn the level of noise emitted.
4. Vibration isolators placed under or around a noise source to prevent vibration from being transmitted to other surfaces, such as floors, walls, or enclosure panels, where additional sound may be generated.

Hearing Protection: Primary consideration should be given to protecting the hearing mechanism. Once a hazard is detected, the initial steps taken should be aimed at hearing protection. Source modification and path modification often require implementation time, whereas steps to protect hearing can be taken immediately. In some instances, this can be accomplished by simply breaking up activity periods or by rotating persons in and out of the hazard area. These procedures increase the intermittency of the noise and thus decrease the threat of damage.

Another means of hearing protection involves the use of personal protective devices or ear protectors (10). These devices usually take the form of ear muffs worn over the external ear so as to provide an acoustical seal against the head, or ear plugs that provide an acoustical seal at the entrance to the external ear canal. The particular type of ear protector worn depends on such factors as the individual's ear anatomy and the environment of the person being protected. Hearing protectors will provide effective hearing protection only if there is an effective hearing conservation program to assure proper fitting, wearing, and motivation at and away from the work place. It should be pointed out that the only unequivocal means for evaluating the effectiveness of personal protectors is to measure the hearing thresholds of the user periodically.

Personal Hearing Protection

Sound attenuation data available from manufacturers of personal hearing protection devices provide a means of calculating the A-weighted sound level reaching the worker's ears. These data are usually in the form of attenuation values for octave bands of noise from 125 to 8,000 Hz. Noise spectra are customarily measured in octave bands to provide identification of noise sources for the purpose of quieting the source. Taken together, these data can be used to calculate the A-weighted sound level reaching the worker's ears.

A simplified diagram of the sound path is shown in Figure 11.1 to help understand this calculation.

Since the noise spectra are measured at the position of the worker's ears, the primary modifier in the sound path is the hearing protection device—the ear muffs or ear plugs being worn. It is best to use a systematic approach to insure the greatest accuracy in this calculation.

First, arrange the data in a table as in Table 11.1. The sound level in each of the octave bands is arranged in columns across the top of the table. The next row in the table is the A-weighting adjustment taken directly from sound level meter specifications. The negative signs indicate an attenuation (loss) and the + signs indicate a gain. In each octave band the levels are combined numerically to give the A-weighted levels in the third row of the table. This row now represents the A-weighted noise spectrum at the worker's ear. The attenuation data for the hearing protector, as supplied by the manufacturer, is shown in the fourth row. Since these are all attenuation (loss) values, they should be subtracted from the A-weighted levels in the area to find the A-weighted levels reaching the worker's ear. The octave band levels in Row 5 represent the A-weighted spectrum at the worker's ear (under the hearing protector). To find a single value of the sound level (as you would expect to read on a sound level meter) for the purpose of determining the noise exposure, these octave band levels in Row 5 can be combined by any of the methods described in Chapter 9. For most purposes, the "shortcut" method with its 1 dB accuracy is sufficient. A systematic approach to

Table 11.1
Calculation of A-Weighted Sound Level at Worker's Ear (data in dB).

Octave band center frequency in Hz	*125*	*250*	*500*	*1000*	*2000*	*4000*	*8000*
Octave band level in area	96	94	96	95	94	96	98
A-weighting	−16	−9	−3	+0	+1	+1	−1
A-weighted level in area	80	85	93	95	95	97	97
Protector attenuation	10	12	20	31	34	40	41
A-weighted level at ear	70	73	73	64	61	57	56

Levels combined by "shortcut" method

76

60

64

67

77

78 dB(A)

using this method is illustrated at the bottom of Table 11.1. This notation helps to insure that the octave band levels are each used only once in the calculation.

Occasionally hearing protector manufacturers supply attenuation values for 3,000 Hz and 6,000 Hz in addition to the octave band data

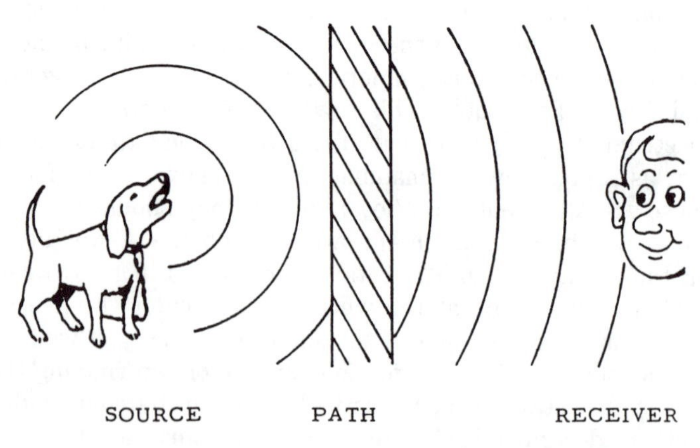

Figure 11.4 Sound transmission.

shown. These figures should be taken into account by combining attenuation data for 3,000 Hz with that for 4,000 Hz and data for 6,000 Hz with that for 8,000 Hz. If the differences in attenuation values are less than 3 dB, the error in using a simple arithmetic average calculation is negligible for this purpose. However, keep in mind that this type of calculation is fundamentally incorrect for logarithmic values such as dB.

Noise Exposure Limits and OSHA

The development of effective and practical requirements and procedures for assuring the health and safety of workers who are exposed to high level noise is very complex. In addition to the very complicated technical aspects related to the effects of exposure to high level noise, the procedures for measuring noise dosage, and the procedures for hearing measuring and impairment assessment, the very important factor of the economic impact on industry must also be considered. The Occupational Safety and Health Administration (OSHA) of the U.S. Department of Labor must face these difficult problems to meet its responsibility in developing and enforcing rules and regulations to limit exposure to potentially harmful noises.

The noise exposure limits set forth by OSHA (14) are designed for both continuous and impulsive noises. The continuous noise limit is set at 90 dB measured with an A-weighting for exposures of eight hours per day, with higher levels being permitted over less time at the rate of an increase of 5 dB for halving of exposure time. For examples, see Figure 11.5. Exposure to continuous A-weighted noise levels greater than 115 dB are not allowed under any circumstances. The limit to impulsive noise exposures is 140 dB peak sound pressure level. (Refer to Figure 9.1 for a more detailed decibel scale.)

When daily noise exposure is composed of two or more periods of continuous exposure at different, steady-state levels, their combined effect is determined by adding the individual contributions as follows:

$$\frac{C_1}{T_1} + \frac{C_2}{T_2} + \frac{C_3}{T_3} + \cdots \frac{C_N}{T_N}. \tag{11.1}$$

The values C_1 to C_N indicate the *time of actual exposure* to specified levels of noise, while the corresponding values of T indicate the total time of exposure *permitted* at each of these levels. If the sum of the individual contributions ($C_1/T_1 + C_2/T_2 + \cdots$) exceeds 1.0, then the mixed exposures are considered to exceed the overall dose limit value. For example, if a worker should be exposed to an L_A of 90 dB for five hours, 100 dB for one hour, and 75 dB for three hours during an eight

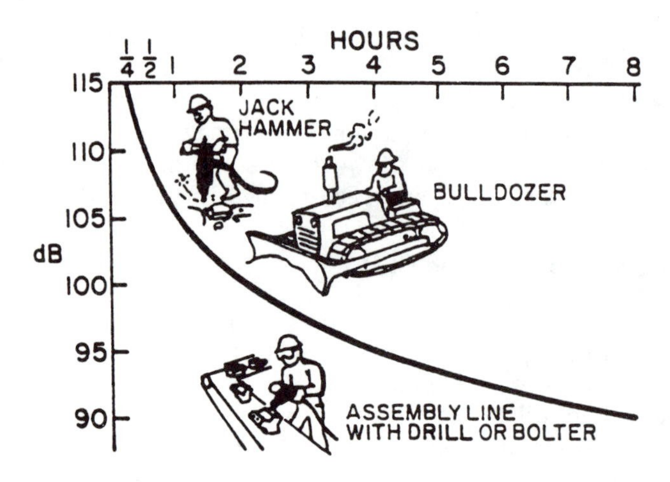

Figure 11.5 OSHA noise exposure limits.

hour working day, then the times of exposure are $C_1 = 5$ hr, $C_2 = 1$ hr, $C_3 = 3$ hr; and the corresponding OSHA limits are $T_1 = 8$ hr, $T_2 = 2$ hr, and $T_3 =$ infinity. Therefore, the combined exposure dose for this worker would be $5/8 + 1/2 + 3/\infty = 1.125$, which exceeds the specified limit of 1.0.

The impulsive noise exposure limit of 140 dB peak sound pressure level of the 1972 OSHA Rules and Regulations (15) does not specify a limit for the number of impulses that a person can be exposed to in an eight-hour working day, but it can be expected that a limit such as 100 impulses for eight hours may be set in a modification of the OSHA noise criteria. Perhaps different peak level limits could be specified for a greater number of impulses, such as 135 dB for 100 to 1,000 impulsive sound exposures; 130 dB for 1,000 to 10,000 impulsive sound exposures; and 125 dB for more than 10,000 impulsive sound exposures in eight hours.

The noise exposure limits specified by OSHA are not intended to provide complete protection for all persons. They are set forth as the most restrictive limits that are deemed feasible with due consideration given to other factors, such as economic impact. Therefore, wherever feasible, hearing conservation measures should be initiated at levels considerably below those specified by OSHA. The ideal action point for initiating hearing conservation measures would be an L_A of about 75 dB for continuous, steady-state noise exposures of eight hours. However, the economic impact of limits set at this low sound pressure level may not be feasible in many situations. Many activities away from the work place cause continuous exposures to noise levels greater than 75 dB (A), so something must be done with the normal lifestyle of people in this country if exposures are to be changed radically. Certainly every effort should be made to institute hearing conservation measures for extended exposures above 80 dB (A).

Lowering noise exposure has very meaningful benefits other than to avoid an OSHA citation. Obviously, the most important benefit is that noise-induced hearing loss may be prevented. In addition, the lower levels will generally afford better working conditions, which should reduce annoyance and improve communication; thus, safety conditions and the general well-being of workers should be improved. Economic advantages of lower noise levels should include increased production and a reduction in compensation claims (in future years) for noise-induced hearing loss. Possibly the OSHA limits for noise exposure may be lowered in the future, so it is generally more economical to have noise levels as low as is feasible now rather than attempt control measures twice.

Other widely used noise exposure limits include those developed by the U.S. Air Force (16), the U.S. Army (17), the Mining Enforcement and Safety Administration (MESA) (18), and the Committee on Hearing, Bioacoustics and Biomechanics (CHABA) of the National Research Council (19).

REFERENCES

1. Peterson, A. G. and E. E. Gross, Jr. Chapter 4 in *Handbook of Noise Measurement*, General Radio Co., Concord, MA, 1974.
2. Ward, W. D. "Effects of Noise on Hearing Thresholds," in W. D. Ward and J. E. Fricke (editors) *Noise as a Public Hearing Hazard*. American Speech and Hearing Association Report 4, 40–48, 1969.
3. Ward, W. D., A. Glorig and D. L. Sklar. "TTS from Octave-Band Noise-Applications to Damage Risk Criteria," *J. Acoust. Soc. Am.* 31:522–528, 1959.
4. Glorig, A., W. D. Ward and J. Nixon. "Damage Risk Criteria on Noise," *Arch. Oto-Laryngol.* 74:413–423, 1961.
5. "Public Health and Welfare Criteria for Noise," U.S. Environmental Protection Agency, EPA Document Number EPA-550/9-73-002, July 1973.
6. Davis, H. "Guide for the Classification and Evaluation of Hearing

Handicap in Relation to the International Aduiometric Zero," *Trans. Am. Acad. Ophth. and Otol.* 69:740–751, 1965.

7. Niemeyer, W. "Speech Discrimination in Noise-Induced Deafness," *Internat. Audiol.* 6:42–47, 1967.
8. Gallo, R. and A. Glorig. "Permanent Threshold Shift Changes Produced by Noise Exposure and Aging," *Amer. Ind. Hyg. Assoc. J.* 25:237–245, 1964.
9. Cohen, A., J. Anticaglia and H. H. Jones. "Sociocusis-Hearing Loss from Non-Occupational Noise Exposure," *Sound and Vibration* 4:12–23, 1970.
10. *Industrial Noise Manual*, American Industrial Hygiene Association, Southfield, MI, 1966.
11. Geiger, P. H. *Noise Reduction Manual*, Engineering Research Institute, The University of Michigan, Ann Arbor, MI, 1953.
12. Hines, W. A. *Noise Control in Industry*, Business Applications Limited, London, 1966.
13. Harris, C. M., ed., *Handbook of Noise Control*, McGraw-Hill, New York, 1957.
14. "Occupational Safety and Health Standards," U.S. Department of Labor, Williams-Steiger Occupational Safety and Health Act of 1970, Federal Register 36, No. 10518, 1971.
15. "Occupational Safety and Health Standards," U.S. Department of Labor, Williams-Steiger Occupational Safety and Health Act of 1970, Federal Register 37, No. 202, 1972.
16. "Hazardous Noise Exposure," Air Force Regulation 160–3, Department of the Air Force, 1964.
17. "Noise Limits for Army Material," Military Standards, MIL-STD-1474 A (MI), Department of Defense, Washington, DC.
18. "Mandatory Health Standards," U.S. Department of Labor, Williams-Steiger Occupational Safety and Health Act of 1970, Federal Register 35, No. 12739, 1971.
19. Kryter, K. D., W. D. Ward, J. D. Miller and D. H. Eldredge. "Hazardous Exposure to Intermittent and Steady State Noise," *J. Acoust. Soc. Am.* 39:451–464, 1966.

CHAPTER 12
WHOLE-BODY VIBRATION

Vibrations in the frequency range of 0.1 Hz to 1,000,000 Hz can be transmitted through solid structures either intentionally or inadvertently. This frequency range extends above and below the range of audible frequencies. While these frequencies can easily be detected by special sensors, they cannot all be heard unless, as we learned in Chapter 4, the vibrating structure is radiating sound waves that are within the frequency range of human hearing.

Vibrations transmitted through a structure are referred to as "structure-borne" vibrations. When a person comes in contact with a vibrating structure these vibrations are then conducted into his body. The vibrations may be applied to particular parts of the body such as the head or limbs or they may be transmitted to the body as a whole from the supporting surface. This latter case is then referred to as "whole-body vibration".

Although vibrations can be introduced in solid materials by sound waves impinging on the surface of a structure, the source of these vibrations is usually some active device mounted on the structure. This chapter is concerned with the effects of vibration on the human body.

Whole-body vibration may have both physiological and psychological effects on the individuals who are exposed to it. These effects depend on many complicated and interrelated factors, such as the magnitude and frequency of the vibration; the location, area, and direction of application; and individual variations in susceptibility. An individual's susceptibility to the effects of vibration is determined by physical state, age, muscle tone, size and weight, etc. The effects of vibration may be heightened or diminished by the physical or mental state of exposed individuals, their activity, or the presence of additional environmental stressors such as concurrent exposure to noise or heat. The vibration frequency largely determines the kinds of effects experienced (1,2). Adverse effects may range from motion sickness (kinetosis), which occurs primarily from exposure to very low frequency vibration (between 0.1 and 1.0 Hz), to local tissue heating and possible cell damage which can result from exposure to vibration with frequencies in the ultrasonic range (above 20,000 Hz).

In this chapter vibrations with frequencies above 1,000 Hz will not be considered because: (*1*) humans are relatively insensitive to these high frequencies, and (*2*) high frequency vibrations are attenuated very rapidly as they propagate away from the source. An International Standard, ISO 2631, "Guide for the Evaluation of Human Exposure to Whole-Body Vibration," sets forth many of the particulars that define and specify the vibration environment (3). According to the vibration perception threshold criteria, outlined in this standard, the descriptive parameters of vibration exposure are: vibration frequency, acceleration magnitude, and the way that the human body is vibrated.

CHARACTERIZATION OF VIBRATION

Spatial Character: Vibration is a vector quantity that may be either rectilinear or angular (rotational). Rectilinear vibration along any one of the three orthogonal axes with respect to the human body will be the primary focus of this discussion.

Temporal Character: Vibration perception criteria normally specify vibration levels that correspond to threshold levels of average or normal individuals in good health. These vibration levels may be either periodic or random in time and may have a distributed frequency spectrum. The duration of vibration exposure is not usually specified in vibration perception criteria; however, the total time of exposure can lead to various biological and/or performance effects.

Magnitude: The quantity used to measure the "amount" or magnitude of vibration may refer to the displacement, velocity, or acceleration. The acceleration magnitude is usually expressed as a root mean square (rms) value in nondimensional units of *g*'s (see Table 12.1) Where one *g* (= 980.665 cm/sec^2) is the value of the standard acceleration due to gravity at the earth's surface. Acceleration magnitude may also be expressed as a level in dB referenced to some standard value, or expressed in peak values as in the graph in Figure 12.1.

Spectral Character: Vibration may occur with many different frequency compositions. Discrete-frequency vibration may consist of a single frequency component or multiple components; distributed-frequency vibration may be composed of a single narrow band of frequencies or a combination of more than one such narrow band of frequencies resulting in broad-band distributed vibration.

Transmission: The transmission of vibrational energy from a source through the ground and/or structures to a reception location may involve

Table 12.1
Vibration Threshold of Perception Criteria

Values are for the root mean square acceleration in units of $g = 980.665$ cm/sec^2 for the frequencies at the center of the 1/3-octave bands beginning 1 Hz and ending 1,000 Hz. Values have been determined from the curves of Figure 12.1.

Center Frequency of 1/3-Octave Band in Hz	*Acceleration (rms) for Threshold of Perception in* g = *980.665 cm/sec*^{2}	
	Whole-Body	*Fingertip*
1	1.8×10^{-3}	....
1.25	$\approx 2.8 \times 10^{-3}$	....
1.6	3.2×10^{-3}	....
2	2.7×10^{-3}	....
2.5	2.3×10^{-3}	....
3.15	2.1×10^{-3}	....
4	1.9×10^{-3}	....
5	1.8×10^{-3}	8.3×10^{-4}
6.3	2.0×10^{-3}	1.2×10^{-3}
8	2.6×10^{-3}	1.9×10^{-3}
10	3.5×10^{-3}	2.7×10^{-3}
12.5	4.4×10^{-3}	3.9×10^{-3}
16	6.0×10^{-3}	5.9×10^{-3}
20	7.7×10^{-3}	8.3×10^{-3}
25	1.0×10^{-2}	9.2×10^{-3}
31.5	1.2×10^{-2}	9.0×10^{-3}
40	1.4×10^{-2}	8.0×10^{-3}
50	1.4×10^{-2}	6.9×10^{-3}
63	....	5.8×10^{-3}
80	....	4.9×10^{-3}
100	....	4.2×10^{-3}
125	....	3.6×10^{-3}
160	....	3.4×10^{-3}
200	....	3.6×10^{-3}
250	....	4.9×10^{-3}
315	....	8.8×10^{-3}
400	....	1.8×10^{-2}
500	....	4.6×10^{-2}
630	....	1.6×10^{-1}
800	....	1.0
1,000	....	(about 8.0)

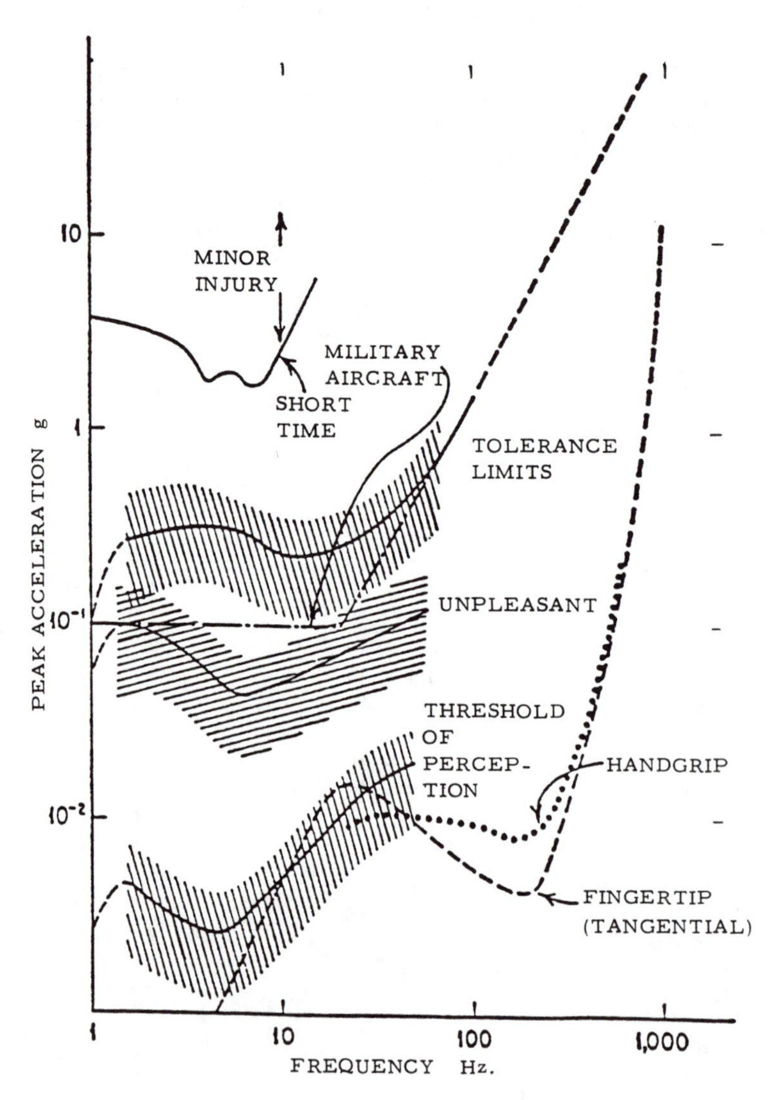

Figure 12.1 Vibration tolerance criteria for exposure time 5 to 20 minutes. Shaded areas represent ± 1 standard deviation. Short time curve is for 1 minute exposure. Tolerance limit for military aircraft is for long-term exposure.

many changes in the characteristics of the vibration along the transmission path. Properties of the transmission medium (or media) and reception structures can be expected to change the magnitude, direction, and frequency spectrum of vibration along its path of propagation. Recall from Chapter 4 that a change in mechanical impedance can result in transmission loss and/or refraction. Of particular note will be relatively large-magnitude vibrations that may be induced at frequencies corresponding to resonant frequencies of receiving structures. Consequently, the description or measurement of vibration must include a detailed description of the point of measurement.

VIBRATION PERCEPTION

Subjective awareness of vibration will depend upon: (*1*) the frequency and the magnitude of the stimulus, (*2*) the individual's response characteristics, and (*3*) the environmental conditions. The vibration perception threshold may be defined as the minimum vibrational acceleration that is necessary to cause a normal person to have a touch (physical contact) or visual sensation of vibration. In some cases an individual may be unaware of levels of vibrations that are higher than those of his threshold of perception because of distracting conditions. However, once attention is directed to the vibration it is almost certain that the individual will detect or become aware of that vibration.

At frequencies below one Hz, vibration is sensed primarily by means of vestibular organs along with the skin surface receptors that may be in contact with the vibrating structure. Above 1.0 Hz where the body resonances and phase shifts occur in the transmission of vibration, the vestibular sensation is augmented by stimulation of mechano-receptors throughout the body. These include receptors in the muscles, tendons and joints as well as in the skin and in the viscera. There may be visual cues as well. The sensations produced by whole-body vibration at frequencies less than 50 Hz vary with frequency and are related to body resonances. Beginning at about 15 Hz, the skin may be considered as the chief sensing mechanism for vibration detection. The threshold of skin surface perception tested at a fingertip is the lowest (most sensitive) in the region of 200–300 Hz. Sensitivity depends on the area, site and pressure of application and is also related to muscle tone.

Many threshold criteria for perception of rectilinear vibration exist in the literature. One set (4) is based on a simple average of results of laboratory experiments involving human perception of single frequency, whole-body vibration in standing, sitting and lying positions. These data, which cover the frequency range of vibration from 1 to 50 Hz, are used in this discussion as the whole-body threshold perception level for any body orientation (standing, sitting, etc.). Consequently it may be necessary to measure vibration in several directions and to determine the vector sum of all components before comparing the exposure level with the perception criterion.

In the frequency range from 50 to 1000 Hz, vibration perception criteria are usually expressed in terms of fingertip sensation levels (5). Vibrations with frequencies higher than 1000 Hz are rarely of concern because these vibrations are rapidly attenuated with distance from the source, and because human perception sensitivity decreases rapidly with increasing vibration frequency. The widely accepted vibration perception criteria for the frequency range from 1 to 1000 Hz are presented in Figure 12.1.

Measurement Methods

Coordinate System: The standard orthogonal coordinate system for whole-body vibration measurements is illustrated in Figure 12.2. Note especially that the origin for this coordinate system is placed at the heart. Careful inspection will show that the three axes are fixed in relation to the body, whether standing, sitting or lying down.

Subjective Detection of Vibration: Vibration can easily be detected when levels of vibration are significantly above perception threshold, however, when vibration is at a level above and yet close to the perception threshold, it may require more attention to confirm its presence. It may be necessary to assume a particular orientation or location of the body to become aware of the vibration. The presence of additional persons may reduce the vibration magnitude since the additional mass could change the vibration characteristics. Depending upon the particular circumstances, vibration may be detected as input to the supporting surfaces of the body (standing, seated, or lying down), cutaneous perception (as with the hands on a table, shelf, etc.) or visual observation of vibrating objects.

It should be a matter of practical consideration that only normal activities be included when determining exposure for vibration levels above perception threshold; thus, for example, vibration of a floor joist that is detectable only through direct touch with the fingertips would not constitute a condition producing "normal" awareness of the vibration. However, if this same vibration is transmitted from joist to floor surface

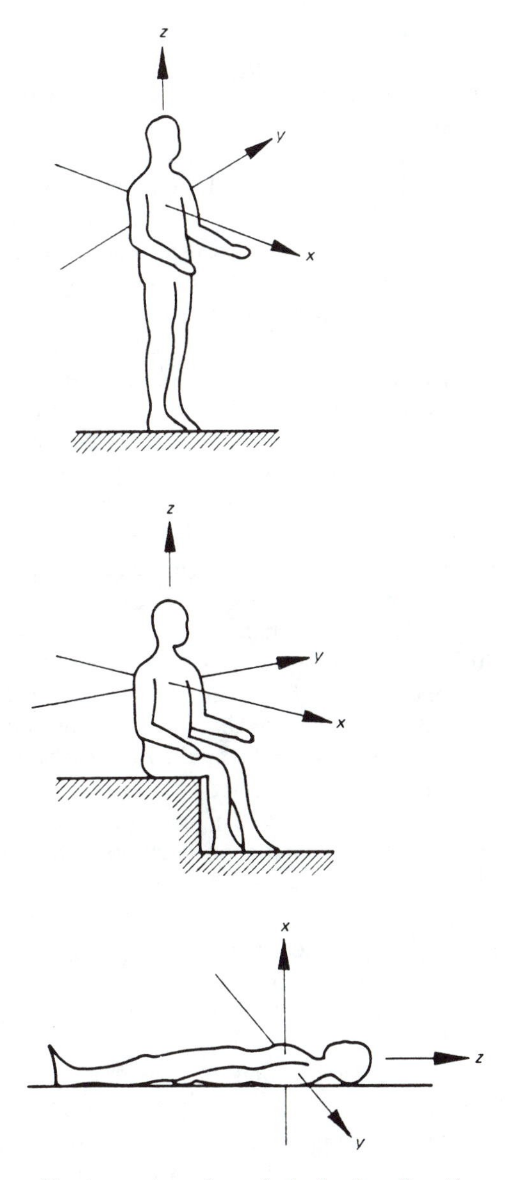

Figure 12.2 Coordinate system for whole-body vibration measurements. The origin is placed at the heart.

x axis = back-to-chest
y axis = right-to-left side
z axis = foot (or buttocks)-to-head

Reproduced with the permission of the International Organization for Standardization, ISO. The complete document, ISO 2631-1974(E) is available from the ISO Central Secretariat and from the ISO member body in the USA: ANSI, 1430 Broadway, New York, NY 10018.

and then detected by a person supported by the floor, the exposure situation may be considered "normal" in determining vibration exposure.

Measurement of Vibration: The root mean square (rms) acceleration levels measured in 1/3-octave bandwidths with center frequencies beginning at one Hz and ending with 800 Hz (which includes all frequencies from 0.9 Hz to 900 Hz) are normally used for the physical measurement of vibration. The necessary equipment for measuring vibration generally includes the following items: a vibration pickup (transducer), a suitable amplifying and signal conditioning device, and a level indicator. More specifically, this system consists of an accelerometer, an amplifier, and a meter with an rms rectifier. Provision for inserting a 1/3-octave band analyzer into the system is often useful. The system should be sensitive to accelerations as low as 0.001 g at frequencies between 1.0 Hz and 10 Hz and as large as 1.0 g at frequencies above 200 Hz. Instruments that meet these requirements are commercially available. In addition, calibrators are available that provide a known vibration (acceleration) level to insure accurate and consistent readings.

Instructions supplied by the manufacturer for the use of the accelerometers should be closely followed (6–9). Accelerometers can be used to measure vibration over wide frequency and dynamic ranges, but particular attention must be paid to the location and mounting of the accelerometer. If possible, the unit should be mounted on a rigid and smooth surface that experiences the vibration to be measured. The axis of the unit will designate the vector component of vibration being measured, and consequently, should be documented. Three-component accelerometers are available that combine three units oriented in mutually orthogonal direction such that the resultant vector may be fully determined from the magnitudes of the orthogonal components. However, a single unit may be used to make separate measurements along the three orthogonal directions or to measure the acceleration magnitude along the major axis of vibration.

The accelerometer may be mounted by any of several methods. Generally, a threaded hole or bolt is provided in the base of the accelerometer that permits mounting the unit (*1*) directly to the surface, (*2*) to a special adaptor that may be cemented to the surface, or (*3*) to a magnetic base that can be attached readily and securely to ferromagnetic materials. The accelerometer may also be attached by means of double-sided adhesive tape, cements (for permanent installations), or greases. In all cases, the mating surfaces should be smooth and free of contamination.

A location of the accelerometer for whole-body measurements should be chosen that corresponds to the point or points where the vibration is experienced by the person with the person in place. Examples are flat surfaces of floors, desk and table tops, chair seats, etc. whereon the accelerometer is mounted directly.

The measuring equipment should be calibrated before and after measurements and at any time during measurements when operation may become suspect, for example, whenever the transducer suffers a severe shock as from a fall. The measured acceleration magnitudes should be compared with established vibration perception thresholds such as those in Table 12.1.

The axis of the measurement should be recorded. In certain cases where the axis does not correspond to the major axis of vibration, each of the three orthogonal components should be measured and evaluated with regard to the perception criteria along with the magnitude of the vector resultant. (The resultant is equal to the square root of the sum of the squares of the orthogonal components.) Whenever a measured level in any 1/3-octave bandwidth exceeds the corresponding threshold of perception, the vibration level may be unpleasant or unacceptable.

REFERENCES

1. Guignard, J. C., Chapter 28, "Introduction" and Chapter 29, "Vibration," in *A Textbook of Aviation Physiology*, edited by J. A. Gillies, Pergamon Press, New York, pp. 807–894, 1965.
2. Goldman, D. E. and H. E. von Gierke, Chapter 44, "Effects of Shock and Vibration on Man," in *Shock and Vibration Handbook*, Vol. 3, edited by C. M. Harris and C. E. Crede, McGraw-Hill Book Company, New York, pp. 44–1 to 44–51, 1961.
3. International Organization for Standardization *Guide for the Evaluation of Human Exposure to Whole-Body Vibration*, ISO Standard, Publication ISO 2631–1974, 1974.
4. Goldman, D. E., "A Review of Subjective Responses to Vibratory Motion of the Human Body in the Frequency Range 1 to 70 Cycles per Second," Report No. 1, Project NM 004 001, Naval Medical Research Institute, National Naval Medical Center, Bethesda, MD, 16 March 1948.
5. von Bekesy, G., "Uber die Vibrationsempfindung," Akust. Z., Vol. 4, pp. 316–334, 1939.
6. *Mechanical Vibration and Shock Measurements*, edited by J. T. Broch, B&K Instruments, Inc., 5111 West 164th Street, Cleveland, OH, Revised Edition, 311 pages, May 1972.

7. *Piezoelectric Accelerometer Manual*, edited by D. Pennington, Endevco Corporation, Pasadena, CA, 119 pages, 1965.
8. *Handbook of Noise Measurement*, edited by A. P. G. Peterson and E. E. Gross, Jr., GenRad, 300 Baker Avenue, Concord, MA, 322 pages, 1972.
9. *Catalog of Instrumentation Specification Sheets*, Wilcoxon Research, Post Office Box 5798, Bethesda, MD.

APPENDIX A.
STUDY GUIDE

On the following pages you will find a set of study questions. These study questions will be used as a basis for examinations and quizzes. While the instructor reserves the right to develop other questions than those presented, the basic format of the enclosed questions replicates quiz and examination material. Students who can answer the questions in this Study Guide correctly will have mastered the essentials of this acoustics course.

CHAPTER 1
MEASUREMENT AND THE LAWS OF MOTION

1. In this unit we have introduced the concept of modeling. What model have we examined? What physical event does it model? Why do we use modeling in this study of physics?
2. List and define the three properties used to describe the motion of an object. Show the basic form of the dimensions of each of these properties.
3. Show graphically the relationships among the three properties of motion for a system moving in a straight line at three different velocities.
4. Define *mass* and *weight* and explain how these two are related.
5. Explain what is meant by the term *force*. If a body has a mass of 2 kg and is moving at an acceleration of 3 m/sec^2 due south, what is the force acting on the body?
6. A 3 kg body has two forces acting upon it. The first force acting alone would move the body northwards at an acceleration of 1 m/sec^2. The second force acting alone would move the body southwards at an acceleration of 3 m/sec^2. What is the resultant force (sum of forces) acting on the body?
7. Define the term *energy* and briefly discuss the three types of energy we have studied in this unit.

8. Define the following terms and tell in each case whether the term refers to a vector or a scalar quantity:

   Distance
   Mass
   Energy
   Velocity
   Speed

9. The graph below shows the distance of a mass from a fixed reference point as a function of time. Describe the sequence of motions the

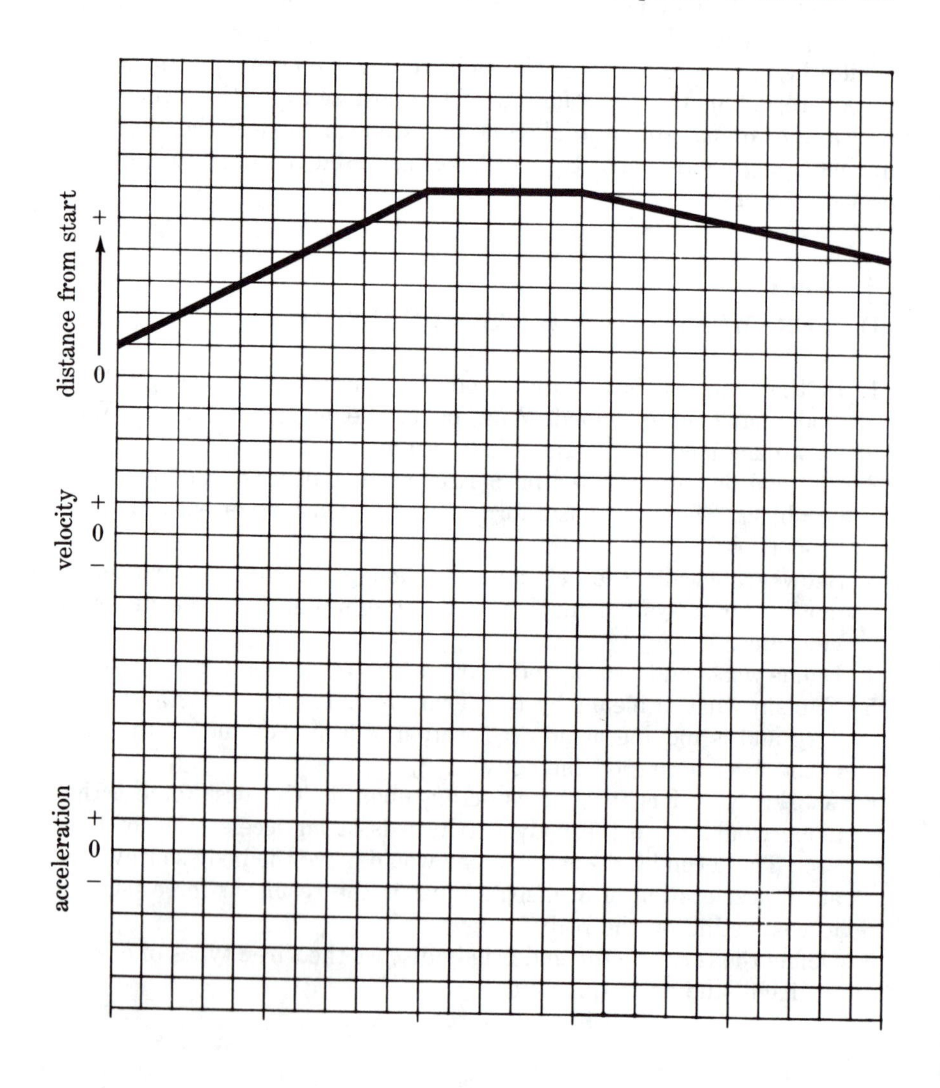

object is undergoing and draw graphs of the velocity and acceleration of the mass over the same time period.

CHAPTER 2
PERIODIC MOTION

1. Define *periodic motion* and *simple harmonic motion*. Do these terms differ in meaning? Explain. Discuss the forces acting on a spring-mass system in motion and tell what physical aspects of the system give rise to these.
2. A pendulum has a swing rate that repeats itself 90 times in one minute. What is the pendulum's frequency in hertz? Briefly discuss the three forms of energy related to pendulum motion and tell how they arise in the pendulum.
3. Show graphically the relationships among the three properties of motion for a system in free simple harmonic motion. How do we describe the relationships between these properties?
4. State the *phase relationships* among the three properties of free simple harmonic motion.
5. List and define the three forms of energy we have examined in our study of simple harmonic motion. Name the *force* related to each of these.
6. Explain what is meant by *damping*, *critical damping* and *decay modulus*. What causes damping in a system and how does it work?
7. What two effects does friction have on the simple harmonic motion of a system? What is the *decay modulus* and to what two system properties is it related?

CHAPTER 3
FORCED PERIODIC MOTION

1. Define *free* and *forced* periodic motion and tell how these two differ. Explain what is meant by the *natural frequency* of a system and discuss why a system has a natural frequency.
2. Give the three vector components of characteristic impedance in acoustic systems and tell how they relate to one another. How are these impedance phenomena affected by the frequency of system operation?

3. The components of characteristic acoustic system impedance are related to specific measurable physical characteristics of the system. Describe these relationships.
4. Show the basic physical dimensions of the characteristic impedance of an acoustic system. What is the name of the unit of acoustic impedance?
5. Explain the relationship between the characteristic impedance of an acoustical system and its frequency response? Show this relationship graphically.
6. What is meant by Q? Show the comparative frequency responses of a system with low Q and one with high Q. Give two different ways by which you could alter a system to increase its Q.
7. Explain the difference between the steady state and the transient responses of a system. If a system takes too long to achieve steady state motion what two different aspects of the system could you alter to reduce this time and how would you alter each one?
8. What is the effect on the resultant system motion when two periodic driving forces are applied to a simple system: (a) when the forces have the same frequency but variable phasing? (b) when the forces are at different frequencies? Explain what is meant by the term *beats* and tell how they arise.
9. Show graphically as functions of frequency the magnitude of the characteristic impedance for an acoustic system. Discuss the relationships of your graphed functions to the force-related phenomena that operate on the system.
10. What is the relationship of damping within a system to the system's frequency response and to its time response? Use graphs to illustrate the relationships, using extreme conditions for each case.

For Reference

$$Q = \frac{f_0}{f_2 - f_1}$$

$$Q = \frac{f_0 M}{2\pi R}$$

CHAPTER 4
WAVE PHENOMENA

1. Explain what is meant by a wave. List the three types (modes of propagation) of acoustic waves we can describe and define each.

2. Discuss briefly what happens when a sound wave traveling in a given medium encounters the boundary of a second medium.
3. List and define the two types of impedances with which we are concerned in acoustics. Explain how these are different.
4. Briefly explain what effects the size of a sound source can have on the radiation of sound by the source.
5. Define *acoustic spherical* and *plane waves* and tell how they differ from one another. What is the *inverse square law* and why does it happen?
6. Define *sound power, sound pressure,* and *sound intensity*. Explain their mathematical relationships.
7. An array of three sound sources are mounted close together in a group suspended from the ceiling so that they are about 4 ft from the floor. Each source alone can produce a sound pressure level of 65 dB (ref. 0.00002 N/m^2). What is the combined sound pressure level of all three sound sources operating simultaneously? What would it be if the array were mounted in the center of a wall? What if they were mounted at the intersection of a wall and the ceiling? What if they were mounted in a corner?
8. Two sound sources that produce sound pressure levels of 75 dB and 78 dB (ref. 0.00002 N/m^2) respectively are mounted close together on a framework suspended 3 ft above the floor of a room. What is the combined sound pressure level of the two sources? How much could this be increased by placing them differently in the room? Where would you place them to get the desired increase and why?
9. A sound wave having a sound intensity level of 70 dB travels from air into hydrogen gas. What is the sound intensity level of the transmitted wave? of the reflected wave?
10. A sound wave having a sound intensity level of 76 dB travels from air into steam. What is the sound intensity level of the transmitted wave? of the reflected wave?
11. A sound wave having a sound intensity level of 67 dB travels from air into oxygen. What is the sound intensity level of the transmitted wave? of the reflected wave?
12. A sound wave having a sound intensity level of 82 dB travels from hydrogen into steam. What is the sound intensity level of the transmitted wave of the reflected wave?
13. A sound wave having a sound intensity level of 73 dB travels from hydrogen into oxygen. What is the sound intensity level of the transmitted wave? of the reflected wave?
14. A sound wave having a sound intensity level of 85 dB travels from oxygen into steam. What is the sound intensity level of the transmitted wave? of the relected wave?

For Reference

Intensity

$$I = \frac{W}{4\pi r^2}$$

$$I = \frac{p^2}{\rho c}$$

Level

$$L_I = 10 \log \frac{I}{I_0}$$

$$L_W = 10 \log \frac{W}{W_0}$$

$$L_P = 20 \log \frac{p}{p_0}$$

Transmission and Reflection

$$\alpha_t = \frac{4(\rho_2 c_2)(\rho_1 c_1)}{(\rho_2 c_2 + \rho_1 c_1)^2}$$

$$\alpha_r = \left(\frac{\rho_2 c_2 - \rho_1 c_1}{\rho_2 c_2 + \rho_1 c_1}\right)^2$$

$$\alpha_t = 1 - \alpha_r$$

Table A.4.1
Characteristic Impedances of Selected Gases

Gas	*Impedance rayls*	*Speed of Sound m/sec*
Air	415	343
Oxygen	450	318
Steam	250	405
Hydrogen	100	1,270
Cochlear Fluid	60,000	1,450
Sea Water	1,480,000	1,500

CHAPTER 5
ANALYSIS OF VIBRATING SYSTEMS

1. Calculate the wavelength of sound propagating in air at frequencies of 20 Hz, 250 Hz, 1000 Hz, 3000 Hz, and 10,000 Hz. What will be the wavelengths of sound at these frequencies propagating in water?

2. Calculate the first two resonance frequencies of a 1 meter long pipe, open to air at both ends.

   f_1 = 171.5 Hz f_2 = 343 Hz

3. Calculate the first two resonance frequencies of a closed tube of pure oxygen that is 6 meters long.

   f_1 = 26.5 Hz f_2 = 53 Hz

4. Calculate the first two resonance frequencies of a closed steam pipe that is 4 meters long.

   f_1 = 50.6 Hz f_2 = 101.3 Hz

5. Calculate the first two resonance frequencies of a 10 centimeter long, closed tube of hydrogen.

   f_1 = 6,350 Hz f_2 = 12,700 Hz

6. Calculate the first two resonance frequencies of an air-filled, 50 centimeter long tube open at only one end.

   f_1 = 171.5 Hz f_2 = 514.2 Hz

7. Calculate the resonance frequency of a bottle having a volume of 1,000 cubic centimeters and a neck that is 1 centimeter long and 0.8 square centimeters in area.

   f_1 = 153 Hz

8. Calculate the first two resonance frequencies of a 2.5 centimeter long ear canal.

   f_1 = 3,430 Hz f_2 = 10,290 Hz

9. A bottle has a volume of 1000 cubic centimeters and a neck that is 1 centimeter long and 0.8 square centimeters in cross-sectional area. What is its resonance frequency when it is half full of water? How does this relate to its resonance frequency when it is empty?

   f_1 (half-full) = 218.4 Hz f_1 (empty) = 154.4 Hz

10. Calculate the first resonance frequency of a drumhead having a density of 2 kg per m^2 that is stretched over a circular frame 1.5 m in diameter with a tension of 60,000 N.

    $f_{01} = 88$ Hz

11. Calculate the first two longitudinal resonance frequencies of a centrally-supported steel bar that is 3 m long with a density of 7,700 kg per m^3 and a Young's modulus of 195,000,000,000 N per m^2.

    $f_1 = 839$ Hz $f_2 = 1{,}678$ Hz

12. Calculate the first two resonance frequencies of a string that is 0.5 m long with a density of 0.2 kg per m and a tension of 20,000 N.

    $f_1 = 316$ Hz $f_2 = 632$ Hz

13. Calculate the first two resonance frequencies of a steel bar, clamped at one end, in longitudinal vibration. The bar is 3 m long and has a density of 7,700 kg per m^3 and a Young's modulus of 195,000,000,000 N per m^2.

    $f_1 = 419$ Hz $f_2 = 1{,}258$ Hz

For Reference

Vibration

$$f = \frac{1}{T}$$

$$\lambda = cT$$

$$f = \frac{c}{\lambda}$$

Plane Waves

$$c = \sqrt{\frac{\gamma p_0}{\rho_0}}$$

String Vibration

$$c = \sqrt{\frac{T}{\delta}}$$

$$f_n = \frac{nc}{2\ell}$$

Table A.5.1.
Environmental Acoustics Data

Model	*Material Property* *Elasticity*	*Density*	$c =$	*Boundary Conditions*	$\lambda =$	$f_n =$
Air Column	γp_0	ρ_0	$\sqrt{\frac{\gamma p_0}{\rho_0}}$	open-open closed-closed	2ℓ	$\frac{nc}{2\ell}$
Air Column	γp_0	ρ_0	$\sqrt{\frac{\lambda p_0}{\rho_0}}$	closed-open	4ℓ	$\frac{(2n-1)c}{4\ell}$
Resonator	γp_0	ρ_0	$\sqrt{\frac{\gamma p_0}{\rho_0}}$	with port	$\frac{2\pi}{\sqrt{S/\ell V}}$	$\frac{c}{2\pi}\sqrt{\frac{S}{\ell V}}$ *
String	T	δ	$\sqrt{\frac{T}{\delta}}$	fixed ends	2ℓ	$\frac{nc}{2\ell}$
Membrane	T	σ	$\sqrt{\frac{T}{\sigma}}$	fixed edges	$\frac{2\pi a}{2.405}$	$\frac{2.405}{2\pi a}\sqrt{\frac{T}{\sigma}}$ *
Bar	Y	ρ	$\sqrt{Y/\rho}$	free-free	2ℓ	$\frac{nc}{2\ell}$
Bar	Y	ρ	$\sqrt{Y/\rho}$	fixed-free	4ℓ	$\frac{(2n-1)c}{4\ell}$

Bar with torsional or transverse waves not given—Complex

Plate not given—Complex

*Inharmonic Series

Note: This appears in the text as Table 5.3

Air Columns

$$f_n = \frac{nc}{2\ell} \qquad \text{Open-Open or Closed-Closed}$$

$$f_n = (2n - 1)\frac{c}{4\ell} \text{ Closed-Open}$$

Bar Vibration (Longitudinal)

$$c = \sqrt{\frac{Y}{\rho}}$$

$$f_n = \frac{nc}{2\ell} \qquad \text{Free-Free}$$

$$f_n = (2n - 1)\frac{c}{4\ell} \text{ Fixed-Free}$$

Helmholtz Resonator

$$f_1 = \frac{c}{2\pi}\sqrt{\frac{S}{\ell V}}$$

Membrane Vibration

$$c = \sqrt{\frac{t}{\sigma}}$$

$$f_{01} = \frac{2.405}{2\pi a}\sqrt{\frac{T}{\sigma}}$$

CHAPTER 6. SPEECH PRODUCTION

1. Describe the components of the human speech production mechanism.
2. What is the *source-filter theory* of speech production? How does it relate to vowel production?
3. Describe the process of generating fricatives and stops.
4. Identify ten of the twelve structures indicated in Figure A.6.1.
5. Identify the indicated structures in Figure A.6.2.
   This is a drawing of the:
   1 ____

CHAPTER 7. HEARING

1. You are seated, quietly dozing during a lecture. Suddenly, a door slamming in the hallway wakes you up. What auditory pathway

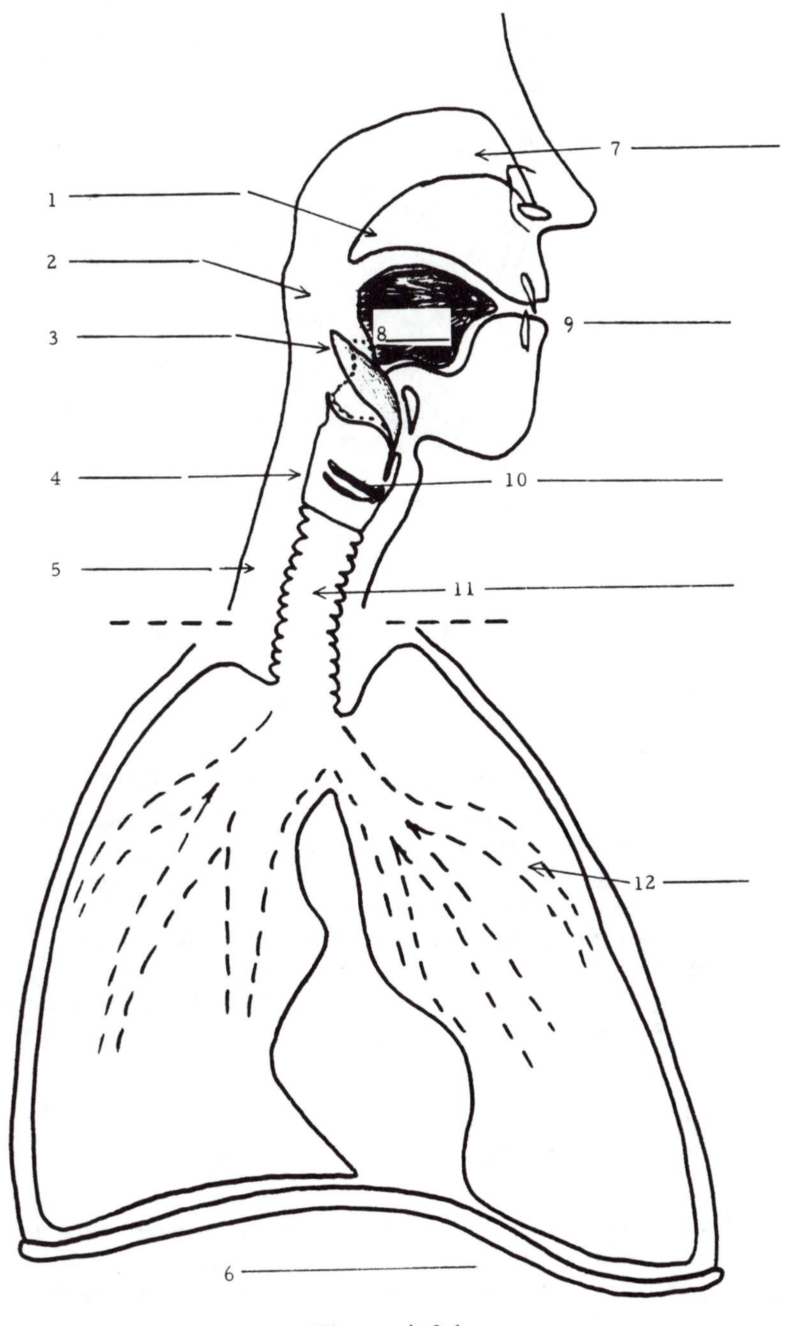

Figure A.6.1.

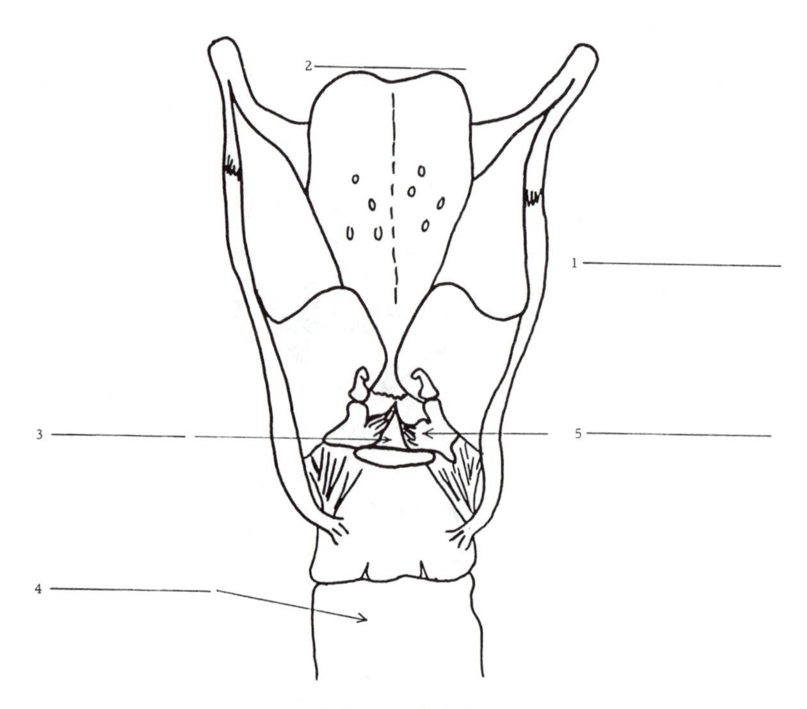

Figure A.6.2.

structures are probably involved in mediating your waking response? Trace the path of this stimulus through the hearing mechanism. What does each structure do in this instance?

2. Describe the function of the middle ear. Discuss normal operation and the response to very loud sounds.
3. Name two functions of the inner ear mechanism. Briefly describe each. Describe the operation of the cochlea.
4. Identify ten of the eleven structures indicated in Figure A.7.1.
5. Identify ten of the eleven structures indicated in Figure A.7.2.

CHAPTER 8
ELECTRO-ACOUSTICS

1. Explain what is meant by *reciprocity* of microphones and loudspeakers. What physical properties make a transducer more suitable for application as a microphone or a loudspeaker? Name one transducer that is not reciprocal.

Figure A.7.1.

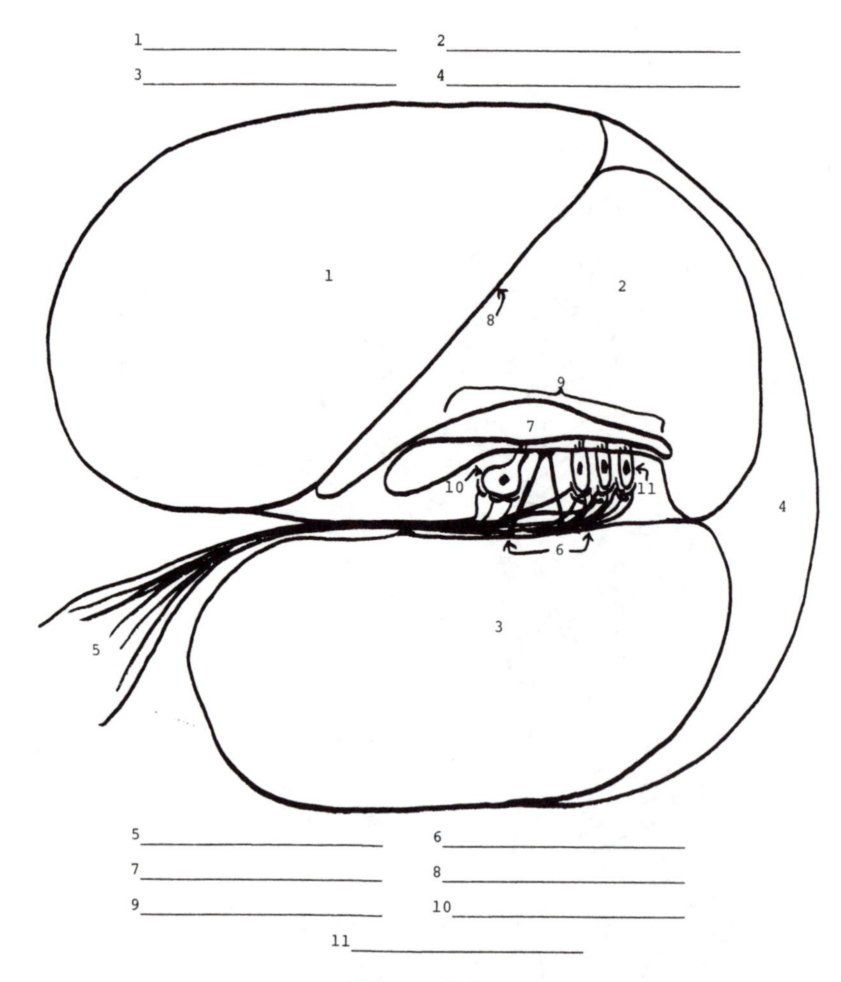

Figure A.7.2.

2. Explain the operation of a *condenser microphone*. Give uses, advantages and disadvantages of this type of microphone. Explain the difference between conventional and electret types.
3. Explain the operation of a *piezoelectric microphone*. Give uses, advantages and disadvantages of this type of microphone. Name two types of piezoelectric microphones and tell how they differ.
4. Compare *dynamic*, *piezoelectric*, and *carbon* microphones. Consider operation, uses, advantages and disadvantages of each type.

CHAPTER 9. SOUND MEASUREMENT

1. Show a block diagram of a sound level meter. Name and describe the function of each component.
2. Name three frequency response networks used in a sound level meter. What is the purpose of including these networks? Under what conditions should they be used?
3. Name two types of microphones used on precision sound level meters. Discuss the differences in their responses. Which one is best suited for measuring moving objects and how should it be oriented with respect to the sound source?
4. Explain three uses for the A-weighting response of a sound level meter. Which network is recommended to be used when calibrating the instrument? How often should the calibration be checked in field operation?
5. Name two microphones with different operating principles that are acceptable for use on sound level meters. Under what conditions should a windscreen be used? Describe the two meter responses used on sound level meters and tell when each should be used.
6. Describe five items of information that should be recorded when making a sound level measurement.
7. A twin turbojet plane is waiting on the runway for take-off. When the first engine is started, the sound power level is 160 dB. Calculate the sound power produced. When both engines are operating, how much is the sound power level increased
8. Normal conversational speech produces a sound pressure level of about 70 dB at the listener's position across the dining table. What is the sound pressure level at the listener's position when two persons are talking? How much is the sound pressure level increased when four persons are talking?
9. A small rock music group can produce an acoustic power of about 0.1 watt. Calculate the sound power level produced. After adding a new power amplifier the group can produce 1.0 watt of sound power. How much has the sound power level changed?
10. In the discussion of the Q of a vibrating system in Chapter 3, the Q factor was defined in terms of the frequencies corresponding to the half-power points on the frequency response curve. Calculate the level in dB of these half-power points referenced to the maximum response of the system.

For Reference

Level

$$L_I = 10 \log \frac{I}{I_0}$$

$$L_W = 10 \log \frac{W}{W_0}$$

$$L_p = 20 \log \frac{p}{p_0}$$

Table A.9.1.
Shortcut method for combining sound levels in dB

Numerical Differences Between Levels	*Amount to be Added to the Higher Level*
0 – 1	3
2 – 4	2
5 – 9	1
> 10	0

Note: This appears in the text as Table 9.3.

CHAPTER 10
ARCHITECTURAL ACOUSTICS

1. Name and discuss three different methods for controlling noise from a stationary source. Tell which aspect of the noise is being affected by each method you name.
2. Name four acoustical design features of the university auditorium (or other civic auditorium). Tell what they are supposed to do and state how well they achieve their purpose. Explain one acoustical improvement you would make in the auditorium.
3. List four acoustical features of your classroom. Evaluate them in regard to the purpose of the room—*teaching*. Give one improvement you would make in this room.
4. You have been assigned to evaluate a set of newly installed therapy rooms. List five specific things you would check in these rooms to insure adequate privacy and speech intelligibility. Explain your choices.
5. A large very noisy copying machine has been installed in an office space. How effective would absorbtive materials be in reducing the

noise at the operator's position? at the reception area 40 ft from the machine? Explain each answer. Give one method other than absorption for reducing the noise from the copier.

6. A classroom is 40 ft by 20 ft and is about 10 ft high. The ceiling is acoustical tile. The walls are painted concrete block and the floor is asphalt tile over concrete. Calculate the reverberation time for this room. Is this acceptable for classroom purposes? If not, how could it be improved?
7. A painted concrete block wall is 20 ft long and 10 ft high. It has a 7 × 3 ft hollow core door and a 3 × 4 ft single glazed window in it. What is the wall's transmission loss?
8. Define the terms: *anechoic chamber*, and *reverberant room*. Tell how each type of room is constructed. What type of sound field is created in each? Give one use of each room.
9. A 4 inch thick brick wall is 10 ft long and 10 ft high. What is its transmission loss with a 7 × 3 ft hollow core door? What if the door were solid core and weather stripped?
10. A painted concrete block wall is 10 ft high and 30 ft long. What is its transmission loss with a 2 × 4 ft single glazed window? What would it be if the window were double glazed with a 1 inch air space? What would it be if the window had a 6 inch air space?
11. A music practice room has painted concrete block walls, an acoustical tile ceiling and a carpeted concrete floor. The room is 10 × 15 × 8 ft. Is the reverberation time appropriate for a music room? If not, how would you change the room?
12. A speech therapy room is 10 × 10 × 10 ft and has painted concrete block walls, a concrete slab ceiling and a carpeted concrete floor. Is the reverberation time appropriate for speech listening? If not, how would you change the room?
13. You have been assigned to do speech therapy in a room with 8 inch thick brick walls and a concrete slab ceiling and floor. The room is equipped with a metal table and chairs. What will be the acoustical problem with this room? List and explain four specific steps you could take to improve the conditions for performing therapy in this room.

For Reference

$$TL = 10 \log 1/\tau$$

$$\tau = \frac{S_1\tau_1 + S_2\tau_2 + \cdots + S_n\tau_n}{S}$$

$$IL_1 - IL_2 = 10 \log_{10} \left(\frac{\alpha_2}{\alpha_1}\right)$$

$$T_{60} = \frac{.161V}{A}$$

$$T_{60} = \frac{.05V}{A}$$

$$A = S_1a_1 + S_2a_2 + \cdots + S_na_n$$

Table A.10.1
Characteristics of Certain Materials.
Data is for 1/3-octave band at 500 Hz.

Material	*Transmission Loss (dB)*	*Transmission Coefficient (τ)*	*Absorption Coefficient* (a)
Walls:			
Wood paneling on stud	30	.001	.17
Plasterboard on stud	38	.0003	.05
Lath and plaster	35	.0003	.04
Concrete block painted	45	.00003	.06
Brick [4 inches thick]	45	.00003	.02
Brick [8 inches thick]	50	.00001	.02
Floor and Ceiling Material			
Acoustical tile	10	.1	.53
Wood floor/plaster ceiling	35	.0003	.09
Concrete slab	40	.0001	.01
Asphalt tile on concrete	40	.0001	.03
Carpet on concrete	40	.0001	.50
Windows:			
Single glazed	20	.01	.18
Double glazed [0.5-inch air]	35	.0003	.18
Double glazed [1.0-inch air]	45	.00003	.18
Double glazed [6.0-inch air]	55	.000003	.18
Doors:			
Hollow core	15	.03	.15
Solid core	20	.01	.10
Weather stripped solid core	25	.003	.10

Note: This appears in the text as Table 10.1

CHAPTER 11
ENVIRONMENTAL ACOUSTICS

1. List and discuss three different effects of high level noise on people.
2. Define the terms: *acoustic trauma*, *noise-induced hearing loss*, *tinnitus*, *temporary threshold shift*, and *permanent threshold shift*.
3. List and discuss four factors (aspects of the noise or listener) that contribute to noise-induced hearing loss.

For the following question refer to Table A.11.1 Environmental Acoustics Data. Assume quiet during lunch hour.

4. A worker spends the time from 8:00 a.m. to 3:00 p.m. (with one hour off for a quiet lunch) in area B of the industrial plant. From 3:00 p.m. to 4:00 p.m. he is in area E and from 4:00 p.m. to 5:00 p.m. he is in area A. Is this worker overexposed to noise? If so, how would you correct the situation?
5. A worker is in area B of an industrial plant from 8:00 p.m. to 12:00 noon. From 1:00 p.m. to 3:00 p.m. he is in area A and from 3:00

Table A.11.1.
Environmental Acoustics Data

Weighting Network Adjustment (rounded to the nearest dB):

Frequency	125	250	500	1,000	2,000	4,000	8,000
A-weighted network	−16	− 9	− 3	+ 0	+ 1	+ 1	− 1
B-weighted network	− 4	− 1	+ 0	+ 0	+ 0	− 1	− 3
C-weighted network	+ 0	+ 0	+ 0	+ 0	+ 0	− 1	− 3

Octave Band Sound Pressure Levels in an Industrial Plant (dB):

Frequency	125	250	500	1,000	2,000	4,000	8,000
Area A	96	89	81	76	73	72	70
Area B	74	78	80	83	82	83	83
Area C	95	98	93	87	85	83	82
Area D	88	92	95	95	94	92	90
Area E	97	96	98	98	99	97	100

Hearing Protector Attenuation (dB):

Frequency	125	250	500	1,000	2,000	3,000	4,000	6,000	8,000
Protector A	12	16	22	34	38	40	46	40	36
Protector B	10	14	23	36	40	44	46	48	44
Protector C	8	10	18	28	34	38	40	40	42
Protector D	20	21	25	36	42	46	44	42	36

p.m. to 5:00 p.m. he is in area C. Is this worker overexposed to noise? If so, how would you correct the situation?

6. A worker is in area A of an industrial plant from 8:00 p.m. to 12:00 noon. From 1:00 p.m. to 2:00 p.m. he is area D. He then goes to area B until 5:00 p.m. Is this worker overexposed to noise? If so, how would you correct the situation?
7. Hearing protector D is being used in industrial area E. What is the A-weighted sound level reaching a worker's ears?
8. Hearing protector C is being used in industrial area B. What is the A-weighted sound level reaching a worker's ears?
9. Hearing protector B is being used in industrial area C. What is the A-weighted sound level reaching a worker's ears?
10. Hearing protector A is being used in industrial area D. What is the A-weighted sound level reaching a worker's ears?

For Reference

$$E = \frac{C_1}{T_2} + \frac{C_2}{T_2} + \cdots + \frac{C_n}{T_n}$$

CHAPTER 12
WHOLE-BODY VIBRATION

1. What acoustical terms are used to describe mechanical vibration? Is vibration considered to be a scalar or vector quantity? In studying the response of the human body to vibration, where is the origin of the coordinate system placed?
2. Name four properties of vibration that determine the effects of vibration on the human body. What are two adverse effects of whole-body vibration that are commonly observed?
3. What are two aspects of vibration that affect human awareness to vibration? What units are generally used to measure vibration? What is the range of frequencies of vibration that produce most of the adverse effects on the human body and what are the effects observed?
4. What sensors are used to measure vibration and how should they be mounted for measurement of whole-body vibration?

APPENDIX B. *EFFECTS OF NOISE ON BEHAVIOR AND WELL-BEING*

The most obvious physiological effect of noise on humans is the temporary and permanent noise-induced hearing losses that may result from high level exposures. However, the effects of noise are not confined to the auditory system. The more subtle effects of noise on behavior and well-being have traditionally been given less attention than the auditory effects, but the annoyance and irritability caused by exposures to noise are striking indicators that noise is a highly potent environmental stressor. The interaction of noise with other environmental stressors has complicated the studies of both physiological and psychological health problems.

ANNOYANCE

Annoyance reactions are perhaps the most widespread response to noise. Annoyance might best be conceptualized as a social response to noise exposure. Noise has often been defined as unwanted sound, and it is this quality that is most often associated with annoyance.

Individual Reactions

It is generally accepted that annoyance increases with sound level, and that higher frequency sounds are more annoying. Also, those sounds that are intermittent or varying over time are rated as more annoying than those that are continuous or unchanging. In addition, annoyance appears to be related to the information content of the sound and the extent to which the sound interferes with some ongoing activity of the individual.

Noise Ratings

Considerable interest has been directed at identifying the measure of noise that best correlates with annoyance. The A-frequency weighting on sound level meters has been, by far, the most widely used frequency weighting applied to community and industrial noise measurements. Both manual and automatic sampling procedures have been used with the A-frequency weighted measurement data.

PHYSIOLOGICAL EFFECTS OF NOISE: STRESS AND HEALTH

This section presents a brief discussion and summary of current knowledge on the non-auditory physiological and health effects of noise. The topics in this section include those that appear most relevant and those that have received the greatest amount of empirical attention. Noise-induced hearing loss is treated in Chapter 11.

The N-Response

The N-Response (1,2) is a group of physiological responses to sound. The response is characterized by:

1. A vasoconstriction of the peripheral blood vessels accompanied by minor changes in blood pressure and heart rate
2. Slow, deep breathing
3. Changes in electrodermal sensitivity, as measured by the galvanic skin response (GSR)
4. A brief change in skeletal muscle tension

These responses cannot be called fear, startle, or anxiety responses because some of them are associated with emotion-arousing activities of the autonomic nervous system, while other are associated with emotion-suppressing activities (1). This pattern of responses begins to appear with noises below an L_A of 70 dB and appears to show adaptation in some cases with repeated stimulation (1).

Circulatory System Effects

Laboratory research provides some evidence that noise affects gross parameters of the circulatory systems, especially for noises above 100 dB. Measures used include blood pressure, pulse rate and heart rate

(3). There is, however, some evidence that working in high noise environments does result in a greater incidence of circulatory problems than working in low noise environments (4). But, as is often the case with the field studies, it is extremely difficult to attribute these effects exclusively to noise as distinct from other stress-producing attributes of the work environment.

Pupillary Dilation

There is evidence, mostly from Europe, that noise affects eye pupil dilation. The magnitude of the effect appears to increase with the intensity of the stimulation, from an L_A of approximately 70 dB to about 110 dB (5). The significance of the response is not known at the present time, but there is an apparent neurological relationship between pupil dilation and the sense of balance (6).

Startle Effects

Startle is a primitive response that may be evoked by a wide variety of stimuli. The purpose of the response is to orient the organism to a potential source of danger. As would be expected, it is particularly susceptible to loud, unexpected noises. The physiological component of the response is essentially independent of the stimulus and includes increased pulse rate, increased blood pressure, and peripheral vasoconstriction. The behavioral component involves a complex pattern of body and facial responses as well as muscular movement. Although the N-response discussed above and the startle response share certain similarities, the patterns are different enough that physiologists consider them to be two different responses (1).

The startle response is normally present at low levels of sound energy, and does tend to show adaptation as a function of repeated stimulation in many, but not all, individuals (7). The significance of these effects is presently unclear.

Vestibular Effects

The vestibular organs of the inner ear (sacculus, utricle, and semicircular canals) are involved in maintaining body balance and orientation in space. The fact that organs important for both hearing and balance exist in such close proximity to each other suggests the possibility of an interrelationship between the two senses. Research has shown that noise can produce dizziness and nystagmus (rapid involuntary side-to-side eye movements). However, in order to obtain these effects, noise levels exceeding an L_A of 130 dB are usually required. Somewhat lower levels, of approximately 120 dB, appear to disturb balance, particularly if the stimulation is unequal at the two ears (8).

At present, there appears to be no evidence that long-term exposure to noise has any significant effect on the vestibular system (9). Further research, however, is warranted.

Stress Reactions

Attempts have been made to explain the effects of noise in terms of physiologic stress theory (10). The theory holds that a large variety of noxious agents are capable of producing a general stress reaction in the organism. Stress is largely non-specific in that different stressors do not each produce a specific set of responses. The organism's response to a stressor is called the general adaptation syndrome (GAS). The GAS has three stages:

1. the alarm stage, in which the system prepares to fend off the stressor;
2. the resistance stage, in which the body fights the stressor; and
3. the exhaustion stage, which occurs if the body can no longer withstand the stressor.

If the stressor is severe enough and present for a prolonged time, the stage of exhaustion is reached and the end result would be the death of the organism from its inability to defend itself against the stressor. In less severe instances, the price is paid in the resistance stage in terms of lowered resistance to infection and the development of the so-called diseases of adaptation—gastrointestinal ulcers, elevated blood pressure, arthritis, etc.

It is fairly well established that noise of extremely high level can act as a stressor and can, at least for some animals, lead to some of the reactions associated with the GAS (11). However, the implications for the human organism are, at present, very unclear. The theory is logically compelling, but the vast complexity and generality of the theory make the determination of the effects of a single stressor such as noise a nearly impossible task. Consideration must be given to the interaction of various stressors, individual differences in susceptibility to stress, and the apparent adaptability of the human organism. Large scale epidemiological and psycholphysiological research is needed.

Figure B.1.

MENTAL HEALTH

Health as defined by the United Nations refers not only to the absence of disease, but to physical, emotional and social well-being (12). Within the scope of this definition, all of the topics covered in this section have some direct or indirect relationship to health. Unfortunately, at the present time, most of these relationships remain undetermined. In fact, very little can be said definitively about the effects of noise on physical or mental health with the exception of its causing hearing loss.

A variety of subjective symptoms such as irritability, anxiety, nervousness, insomnia, loss of appetite, etc. have been associated with noise, but the subjective nature of these effects makes their verification difficult. Field research in noise is often impeded by the difficulty of separating those effects attributable to noise from the effects of other stress-producing stimuli in the environment.

The research reviewed in this section suggests that noise does affect a number of physiological systems of the individual but data are not available to determine if these effects are of a major consequence to health.

Noise has been accused of adversely affecting mental health. For example, recent data suggest a positive relationship between aircraft noise and mental hospital admissions (13). Unfortunately, the criticism of the study was so intense that no valid conclusions concerning noise and mental health can be derived. There is a serious and immediate need for well-controlled, large scale epidemiological research in this area.

Task Performance

Several comprehensive reviews of the effects of noise on task performance have been written (1,14,15). There seems to be general agreement among these reviewers that the research to date has failed to yield a consistent pattern of effects. Noise has been shown to improve task performance, to impair task performance, and, in some instances, to have no apparent effect. Overall, it is probably safe to conclude from these reviews that the effects of noise on short-term task performance are not severe in most cases, and that the detection of these decrements requires detailed performance assessment and the use of noise-sensitive tasks.

In a literature review compiled by the EPA, the following conclusions pertaining to task performance were advanced (6).

- Continuous noise without special meaning does not generally impair performance unless the sound exposure level L_A exceeds 90 dB. Even at this level the effects are not consistent.
- Intermittent and impulsive noises are more disruptive than steady-state noises of the same level. Sometimes levels below 90 dB will produce effects, especially if the bursts come at irregular intervals.

Figure B.2.

- High-frequency components of noise (above approximately 2,000 Hz) usually produce more interference with performance than low-frequency components of noise.
- Noise usually does not affect the overall rate of work, but may increase the variability of the work rate.
- Noise is more likely to increase error rates as opposed to rate of work.
- Complex or demanding tasks are more subject to noise-related impairments than simple tasks.

Characteristics of the Noise, Task, and Individual

The previous conclusions suggest that the effects of noise on performance are related to the nature of the noise, the nature of the task, and the state of the individual.

Distracting or "attention-demanding" noises, such as impulse or irregular intermittent noises or very intense noises, result in greater task interference.

Most performance decrements have been found on tasks that require continuous performance, prolonged vigilance, or the performance of two tasks simultaneously. Tasks that require simple, repetitive operations are unaffected and sometimes enhanced by certain low level noises. Obviously, tasks that require the operator to attend to auditory cues for successful performance are almost always impaired in the presence of noise.

Noise-sensitive tasks, such as those requiring continuous performance or prolonged vigilance, prevent the individual from pacing his performance and penalize the individual for momentary lapses of attention. On the other hand, simultaneous tasks bring about decrements because they overload the information-processing capacity of the individual. The individual has a limited capacity for information processing and where noise is present, less spare capacity exists for task information, relative to quiet conditions (16–18). Consequently, noise-related impairments are often found in overloading or demanding task situations.

In the presence of an arousal-increasing stimulus such as noise, performance on simple or boring tasks might be improved because arousal level is increased toward an optimal level (19). Similarly, the presence of noise during the completion of a difficult or demanding task might result in a supra-optimal level of arousal and impaired performance. Tasks of moderate difficulty would probably remain unaffected by noise.

There appears to be a great amount of variation in the way in which different individuals respond to noise, and although this is a common observation, very little is known about the nature of these differences. There has, however, been an attempt to apply the theory of arousal to the problem of individual differences (20,21). The basic supposition of this approach is that individuals differ in their chronic levels of arousal. If one individual is chronically more aroused than another, an additional arousal afforded by the presence of noise would be more likely to lead to a condition of overarousal for this individual than for a less chronically aroused individual. There is evidence linking the personality dimension of introversion-extroversion to autonomic indices of arousal and performance. It appears as if introverts are more chronically aroused than extroverts. Data are available which suggest that introverts perform better than extroverts in boring and monotonous task situations, and that introverts appear to be more adversely affected by noise than extroverts. These findings must be considered as only tentative, but this does appear to be a promising avenue for future research.

Cumulative and Post-Noise Effects

Research has been conducted which indicates that the adverse effects of noise tend to appear toward the end of task performance sessions (22). This effect appears to increase in magnitude as time spent in noise increases (23).

Recent studies have shown that, although noise may not affect performance during the actual exposure, it may produce impairments which occur after the noise has been noticed on tasks involving proofreading and frustration tolerance. Apparently noise exposure can cause some type of residual or depletion effects. Also, more severe aftereffects were found with irregular-intermittent and intense ($L_A \geqslant 108$ dB) noises, with intermittency or unpredictability of the noise being more important variables than sound level.

These same researchers also found that when subjects were provided with the means to terminate the noise they were exposed to, the magnitude of the post-noise effect was reduced even when this control was not exercised.

Field Studies

Industrial and other work situations do not readily lend themselves to controlled experimentation. As a consequence, much of the previous field research has been subject to severe methodological deficiencies (1). It is usually difficult to separate the effects attributable to noise from those related to other (physical) stressors such as heat and air pollution, or to considerations of accident threat and job security. Evaluation of the positive effects of noise reduction efforts are often confounded by positive morale and motivation changes that also accompany the intervention in the work environment.

A five-year study of medical, attendance, and accident files for 1,000 factory workers showed that workers in high noise settings ($L_A \geqslant 95$ dB) had more job-related accidents, sickness, and absenteeism than their counterparts in more quiet settings ($L_A \leqslant 80$ dB) (24). These results, too, are subject to criticism because it is quite possible that high noise levels are found in work situations that differ in some important respect, such as accident hazard, from those situations with lower noise levels.

Implications of Task Performance Effects

Assessment of the effects of noise on task performance requires consideration of the particular noise involved, the type of task in question, and the individuals performing the task. In general, overall rate of work is not affected, but variability is often increased. Demanding tasks or tasks that must be performed for relatively long periods of time are more subject to disruption by noise. Although in some situations, performance during noise is unaffected, subsequent performance or behavior sometimes suffers as a result of previous noise exposure. Most of these conclusions are based on the results of short-term laboratory research. Valid field research is seriously lacking.

SLEEP DISTURBANCE

There are two aspects to the problem of sleep disturbance: one concerns actual arousal or waking due to the noise, and the other concerns changes within the sleeping individual who does not awaken with the noise.

Stages of Sleep

During the course of sleep the individual typically goes through a progression of different stages of sleep. There are four principal stages, and these have been differentiated through the examination of brainwave activity. With relaxation, the rapid irregular waves change to a regular pattern. Stage 1 follows this period of relaxation, and it is characterized by a prolonged reduction in wave amplitude and frequency. Later, bursts of waves and large slow waves occur. This is Stage 2. Approximately thirty to forty-five minutes later, bursts of high amplitude slow

Figure B.3.

waves commence. This is Stage 3. When these waves are present for 50% of the time, the deepest sleep stage, Stage 4, is entered. After approximately sixty to ninety minutes, brain wave activity again resembles that found in Stage 1. However, it is accompanied by rapid eye movement (REM). This is the REM stage, the stage where most dreaming takes place. It is usually thought that all stages of sleep are necessary for adequate functioning.

Variables Related to Sleep Disturbance

The major variables that appear related to response to noise during sleep are age, sex, stage, noise level, rate of noise occurrence, noise quality, and presleep activity (25,26).

Age: Middle-aged and older subjects are more affected by noise than children and young persons at all stages of sleep.
Sex: Women are typically more sensitive to noise during sleep than men. Middle-aged women are especially sensitive to subsonic jet aircraft fly-overs and simulated sonic booms.
Stage of Sleep: People tend to be most responsive to noise during sleep Stage 1, next during 2, and least during REM and Stage 3 and 4 sleep. Often, in the deeper sleep stages, noise does not produce behavioral awakening, but does result in shifts in stage. Usually, the shift is from a deep to a light sleep. The meaning of

the stimulus is also important, in that more meaningful stimuli elicit greater response. In general, behavioral awakening is more likely to occur the longer someone has been sleeping.

Noise Level: As a general rule, the higher the noise level, the greater the probability of response, regardless of whether the response is defined as awakening or change in sleep stage (27).

Rate of Noise Occurrence: Research to date has yielded conflicting findings in this area. For example, low density traffic sounds have been shown to be more disruptive of sleep than high density sounds, while on the other hand, jet take-offs were found to be as disruptive at low rates as at higher rates (28). It is possible that the background noise levels, the uncertainty and the novelty of the sounds play important roles in sleep disturbance.

Noise Quality: Meaningful sounds awaken an individual at intensities lower than those required for meaningless or neutral sounds.

Presleep Activity: What research there is suggests that presleep activity such as exercise is not closely related to noise sensitivity during sleep. On the other hand, sleep deprivation does seem to increase the amount of time spent in Stages 3 and 4 sleep and REM and consequently should affect noise sensitivity.

Implications of the Sleep Disturbance Effects

Sleeping in noisy environments appears to produce adverse effects either in the form of awakening the sleeper, or in the form of shifts in the stages of sleep. It should be pointed out that the existing data come almost exclusively from laboratory studies employing relatively few participants. There does appear to be a relationship between sleep disturbance and annoyance. Community noise surveys have shown sleep disturbance to be a major source of annoyance (29).

Overall, very little is known about the long-term effects of sleep disturbance. The body needs sleep for normal functioning, and it is quite possible that sleep disturbance will yield adverse health effects. This is especially so for those individuals, such as the elderly, who are most sensitive to noise.

SPEECH INTERFERENCE

Most people have experienced situations in which noise has prevented them from understanding someone's speech, or where they themselves were misunderstood. However, little scientific effort was directed to studying this problem until the advent of the telephone and the development of mechanized military systems. To date, a considerable amount of laboratory research has been done, and much is known about how a given speech sound will be masked by a particular noise (1,30,31). Speech interference is usually considered as one aspect of the general phenomenon of masking. Masking refers to the effect one sound has of making another sound more difficult to hear. One sound may alter the loudness, perceived quality, or apparent location of another sound.

This specialized laboratory research on masking has had limited applicability to the problem of ordinary speech. Ordinary speech is a complicated sequence of sounds with constantly varying level and spectral distributions. Also, for speech to be intelligible it is not necessary that all the sounds be heard. Speech is so redundant, and the typical listener so familiar with the language, that information can be missed and the speech will still be understood.

Variables Related to Degree of Speech Interference

There are a number of variables that influence the extent to which noise will interfere with speech. These are the characteristics of the speaker and listener, the characteristics of the message, and the characteristics of the masking noise.

Characteristics of the Speaker and the Listener: Noise will tend to interfere with speech reception to a greater extent if the speaker has poor articulation, or if the listener and speaker use different dialects. Lack of extensive knowledge of, and experience with, the language will render communication more difficult in noise. Both in terms of their poor articulation and lower degree of language familiarity, children appear to suffer more from background noise than do adults with normal hearing sensitivity. There is tentative evidence that suggests that noise in the home environment may be related to impaired auditory discrimination and reading achievement in children (32, 33). Decrements in hearing acuity due to the aging process (presbycusis) also require lower background noise levels for adequate speech communication (34).

Characteristics of the Message: Research has demonstrated that the intelligibility of speech in noise is related to the probability of occurrence of a given sound, word, or phrase (35). In other words, communications that contain simple and predictable information are less subject to interference from noise.

Characteristics of the Noise: As a general rule, the more intense the noise, the greater will be its interference with speech. The frequency spectrum of the noise is also very important in that the extent to which a given noise will interfere with speech depends in part on the sound pressure levels of the noise at the speech frequencies.

The effect of intermittent or impulse noise on speech intelligibility is difficult to assess. The severity of the effect depends on the frequency and duration of the bursts. As the frequency and duration increase, the level of speech intelligibility is reduced. Infrequent bursts of short duration usually do not interfere with speech in that some information can be missed without making the communication unintelligible.

Measures of Speech Interference

Various schemes have been developed to characterize noise in respect to its speech-masking abilities. The two best known are the Articulation Index (AI) and the Speech Interference Level (SIL) (36). These measures and their variants allow the user the predict the intelligibility of speech of a given level in a specific noise. The AI is the more complicated of the two measures because it takes into consideration the fact that certain frequencies in the noise are most effective in masking the other frequencies. The SIL provides only a measure of the averaged general masking capability of the noise with the lowest and highest frequencies ignored.

The simple A-weighted sound level (L_A in dB) is also a useful index of the masking ability of a noise. The A-weighting process emphasizes mid-range frequencies, as does the SIL. They differ in that the SIL ignores the lower frequencies, where the A-weighting includes them at a reduced level. The choice as to which measure to use depends on the level of accuracy required. The AI is the most accurate, but it is also the most complicated to use (37). In most instances, L_A or SIL measurements are adequate.

Noise Level, Vocal Effort, and Distance

Attempts have been made to graphically portray the dependence of intelligibility on distance between speaker and listener with respect to noise level (38). Figure B.4 shows the distances over which speech can be understood for various noise levels. For example, at three feet a raised voice can be understood through an L_A of 71 dB. By *understood* it is meant that 95% of the key words in the group of sentences will be comprehended. It should be pointed out that these figures apply only

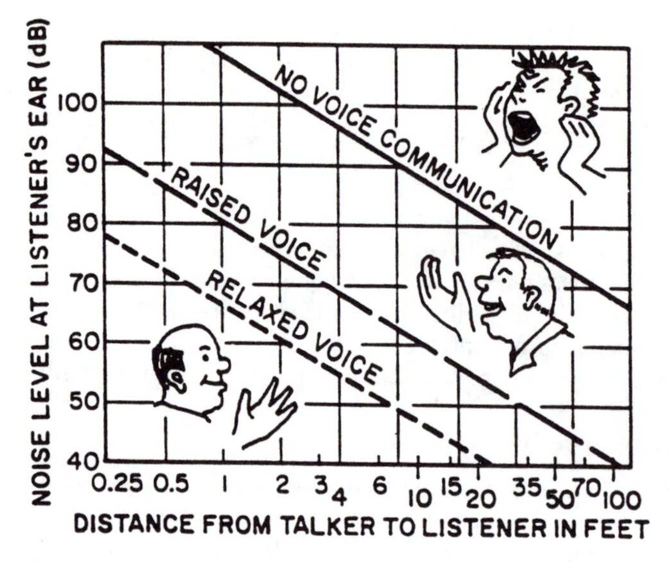

Figure B.4. Relation between noise level and voice communication.

to outdoor environments. Predictions for indoor environments would be more complex because consideration would have to be given to the reverberant qualities of indoor spaces.

Implications of Speech Interference

Noise does interfere with speech. Research on community noise indicates that speech interference is a primary source of noise-related annoyance. In certain situations noise may mask signals that, if not heard, could lead to property damage, personal injury, or even death. Although people can adapt to even relatively high levels of background noise, there is evidence that they develop "noncommunicating" life styles in such environments (39), and this is undesirable in terms of the quality of life. There is also tentative evidence which suggests that noise in the home can adversely affect the language development of children.

SUMMARY

Permanent noise-induced hearing loss is obviously the best documented and most significant effect of exposure to noise. In addition, however, noise has been shown to detract in many ways from the quality of life in our society. It has been demonstrated that, under certain circumstances, noise can produce annoyance, interfere with speech communication, disturb sleep, and disrupt task performance. Noise is also capable of eliciting a variety of physiological responses.

REFERENCES

1. Kryter, K. D. *The Effects of Noise on Man*, Academic Press, New York, 1970.
2. Davis, R. G., A. M. Buchwald and R. W. Frankman. "Autonomic and Muscular Responses and Their Relation to Simple Stimuli," *Psychological Monographs* 69, no. 405, 1955.
3. Etholm, B. and K. E. Egenberg. "The Influence of Noise on Some Circulatory Functions," *Acta Oto-Laryngol.* 58:508–213, 1966.
4. Jansen, G. "Adverse Effects of Noise on Iron and Steel Workers," *Stahl. Eisen.* 81:217–220, 1961.
5. Jansen, G. "Relation between Temporary Threshold Shift and Peripheral Circulatory Effects of Sound," *Physiological Effects of Noise*, B. L. Welsh and A. S. Welsh (editors), Plenum Press, New York, 1970.
6. "Public Health and Welfare Criteria for Noise," U.S. Environmental Protection Agency, EPA Document Number EPA-550/9-73-002, 1973.
7. Landis, C. and W. A. Hunt. *The Startle Pattern*, Farrar and Rinehart, Inc., New York, 1968.
8. Nixon, C. W. "Some Effects of Noise on Man," U.S. Air Force, Report No. AMRL-TR-71-53, Wright-Patterson AFB, OH, 1971.
9. Dieroff, H. G. and H. J. Scholtz. "Zur Frage de Lambedingten Vestibularissghaden bei begutachteten Larmarbeitern," *Z. Largngol. Rhirol. Otol.* 46:746–757, 1967.
10. Selye, H. *The Stress of Life*, McGraw-Hill Book Company, New York, 1956.
11. Welch, B. L. and A. S. Welch, eds. *Physiological Effects of Noise*, Plenum Press, New York, 1970.
12. *The First Ten Years of the World Health Organization*, World Health Organization, Geneva, 1958.
13. Abey-Wickrama, I., M. F. Albrook, F. E. G. Gattoni and C. F. Herridge. "Mental-Hospital Admissions and Aircraft Noise," *Lancet* 297:1275–1278, 1968.
14. Broadbent, D. E. "Effects of Noise on Behavior," *Handbook of Noise Control*, C. M. Harris (editor), McGraw-Hill Book Company, New York, 1957.
15. Gulian, E. "Psychological Consequences of Exposure to Noise: Facts

and Explanations," *Proceedings of the International Congress on Noise as a Public Health Problem*, W. D. Ward (editor), pp. 363–378 (U. E. EPA No. 550/9-73-008), 1973.

16. Hockey, G. R. J. "Effect of Loud Noise on Attentional Selectivity," *Quarterly Journal of Experimental Psychology* 22:28–36, 1970.
17. Hockey, G. R. J. "Signal Probability and Spatial Location as Possible Basis for Increased Selectivity in Noise," *Quarterly Journal of Experimental Psychology* 22:37–42, 1970.
18. Kahnerman, D. *Attention and Effort*, Prentice-Hall, Englewood Cliffs, NJ, 1973.
19. Hebb, D. O. "Drives and the C. N. S.," *Psychological Review* 62:243–254, 1956.
20. Welford, A. T. "Stress and Achievement," *Austrialian Journal of Psychology* 17:1–11, 1965.
21. Welford, A. T. *Fundamentals of Skill*, Methuen, London, 1968.
22. Broadbent, D. E. *Decision and Stress*, Academic Press, London, 1971.
23. Hartley, L. R. "Effects of Prior Noise or Prior Performance on Serial Reaction," *Journal of Exp. Psych.* 101:255–261, 1973.
24. Cohen, A. "Industrial Noise and Medical Absence and Accident Record Data on Exposed Workers," *Proceedings of the International Congress on Noise as a Public Health Problem*, W. D. Ward (editor), pp. 441–454 (U.S. EPA No. 550/9-73-008), 1973.
25. Williams, H. L. "Effects of Noise on Sleep: A Review," *Proceedings of the International Congress on Noise as a Public Health Problem*, W. D. Ward (editor), pp. 501–511, U.S. EPA No. 550/9-73-008, 1973.
26. Lukas, J. S. "Awakening Effects of Simulated Sonic Booms and Aircraft Noise on Men and Women," *Journal of Sound and Vibration* 20:457–466, 1972.
27. Thieseen, G. J. "Effects of Noise from Passing Trucks on Sleep," Report Q1 presented at 77th Meeting, Acoustical Society of America, Philadelphia, PA, 1969.
28. Schieber, J. P. "Etude analytique en laboratoire de l'influence du bruit sur le sommeil," Centre d/Etudes Bioclimatiques du CNRS, Strasbourg, France, 1968.
29. Lukas, J. S. "Measures of Noise Level: Their Relative Accuracy in Predicting Objective and Subjective Responses to Noise during Sleep," U.S. EPA No. 600/1-77-010), 1977.
30. Jeffress, L. A. "Masking," *Foundations of Modern Auditory Theory, vol. 1*, J. V. Tobias (editor), Academic Press, New York, 1970.
31. Webster, J. C. "Effects of Noise on Speech Intelligibility," *Noise*

as a Public Health Hazard, W. D. Ward and J. E. Fricke (editors), ASHA Report No. 4, American Speech and Hearing Association, Washington, DC, 1969.

32. Cohen, S., D. C. Glass and J. E. Singer. "Apartment Noise, Auditory Discrimination and Reading Ability in Children," *Journal of Experimental Social Psychology* 9:407–409, 1973.
33. Wachs, T. D., I. C. Uzgiris and J. McV. Hung. "Cognitive Development in Infants of Different Age Levels and from Different Environmental Backgrounds: An Explanatory Investigation," *Merrill-Palmer Quarterly* 17:283–317, 1971.
34. Palva, A. and K. Jokinen. "Presbyacusis: V. Filtered Speech Test," *Acta Oto-Laryngol.* 70:232–241, 1970.
35. Miller, G. A., G. A. Heise and W. Lichten. "The Intelligibility of Speech as a Function of the Context of the Test Material," *Journal of Experimental Psychology* 41:329–335, 1951.
36. Beranek, L. L. "The Design of Speech Communication Systems," *Proceedings Institute of Radio Engineers* 35:880–890, 1947.
37. Klumpp, R. G. and J. C. Webster. "Physical Measurements of Equally Speech-Interferring Navy Noises," *Journal of the Acoustical Society of America* 35:1328–1338, 1963.
38. Beranek, L. L. "Noise Control in Office and Factory Spaces," *15th Annual Meeting Chem. Eng. Conf. Trans.* Bulletin 18, pp. 26–33, 1950.
39. "Effects of Noise on People," U.S. Environmental Protection Agency, (NTID 300.7), 1971.

Appendix C
GLOSSARY OF SYMBOLS, ACOUSTICAL TERMS, PHONOLOGICAL TERMS, AND AUDIOLOGICAL TERMS

SYMBOLS

A	ampere
A	area of the surface of a sphere
a	acceleration
C	actual duration of exposure to noise
C	Celsius (temperature)
c	centimeter
c	velocity of sound
cm	centimeter
D	directionality of the source
d	distance
E	calculated noise exposure
e	base of natural logarithms = 2.7182818
F	force
f	frequency
f_0	resonance frequency
ft.	foot or feet
G	gravitational constant = 980.665 cm/s^2
gm	gram
Hz	hertz
I	current
I	sound intensity
I_0	reference intensity
J	joule
K	stiffness
k	kilo
kg	kilogram

L	level in decibels (dB)
L_{eq}	energy equivalent level
ℓ	length
M	mass
m	meter
N	newton
n	integer 1, 2, 3, 4, · · · · n
Pa	pascal; the unit of pressure = N/m^2
p	10^{-12}
p	sound pressure
Q	quality factor (efficiency)
R	mechanical resistance (friction)
R	resistance (electrical)
r	radius or distance from a source
r	rate of motion
rms	square root of the mean of the squares
S	area
s	second
T	period
T	permitted duration of noise exposure
TL	transmission loss (in dB)
t	time
V	volt
v	velocity
v	volume
W	sound power
W	watt
W	work
W_0	reference power
Z	mechanical impedance

GREEK SYMBOLS

α	area density mass/unit area
β	$2\pi f_0$
δ	linear density mass/unit length
λ	wavelength
μ	10^{-6}
ν	elasticity
π	3.1415926
ρ	density
ρ_0	volume density

τ decay modulus
τ time since start of observation
τ transmission coefficient

ACOUSTICAL TERMS

absorption coefficient
The ratio of the sound energy absorbed by a surface to the sound energy incident upon that surface.

acoustics
A science that covers the generation, reception, application, and effects of sound.

acoustic intensity
See **sound intensity.**

acoustic power
See **sound power.**

acoustic pressure
See **sound pressure.**

ambient noise
The overall composite of sounds in a given environment.

amplitude
The quantity of sound produced at a given location away from the source or the overall ability of the source to emit sound. The amount of sound at a location away from the source is generally described by the sound pressure or sound intensity, while the ability of the source to produce sound is described by the sound power of the source.

anechoic room
A room that has essentially no boundaries to reflect sound energy generated therein. Thus, any sound field generated within an anechoic room is referred to as a free field.

audiogram
A graphic recording of hearing levels referenced to a statistically normal sound pressure level as a function of frequency.

audiometer
An instrument for measuring hearing sensitivity.

continuous spectrum

A spectrum in which energy components are distributed over a range of frequencies such that some finite amount of energy can be found at any frequency within the range.

critical band

A bandwidth within a continuous-spectrum noise that has a sound power equal to that of a single-frequency tone centered in the critical band and just audible in the presence of the critical bandwidth of noise.

cycle

A complete sequence of unique values of a function that recur in the same sequence at regular intervals.

cycle per second (cps)

See **frequency.**

decibel (dB)

A convenient means for describing the logarithmic level of sound intensity, sound power or sound pressure above arbitrarily chosen reference values.

diffuse sound field

A field characterized by sound pressure levels that are essentially the same throughout in which the directional incidence of energy flux is randomly distributed.

effective sound pressure

The sound pressure found at a given location by calculating the root mean square value of the instantaneous sound pressures measured over a period of time at that location.

free field

A homogeneous isotropic medium free from boundaries. In a free field, sound radiating from a source can be measured accurately without influence from the test space. True free-field conditions are rarely found except in expensive anechoic (echo-free) test chambers; however, approximate free-field conditions exist in any homogeneous space where the distance from reflecting surfaces to the measuring location is much greater than the wavelengths of the sound being measured.

frequency

The rate at which complete cycles of high and low pressure regions are produced by the sound source. The unit of frequency is the cycle per second (cps) or preferably the hertz (Hz). The frequency range of the human ear is highly dependent upon the individual and the sound level, but an ear with normal hearing will have a frequency range of approximately 20 to 20,000 Hz at moderate sound levels. The frequency of a propagated sound wave heard by a listener will be the same as the frequency of the vibrating source if the distance between the source and the listener remains constant; however, the frequency detected by a listener will increase or decrease as the distance from the source is decreasing or increasing (Doppler effect).

hertz (Hz)
See **frequency.**

infrasonic frequency
Sounds which are below the audible frequency range.

intensity
See **sound intensity.**

level
The amount by which a quantity deviates above or below a reference quantity expressed as the logarithm of the ratio of that quantity to a reference quantity expressed in the same units.

loudness
An observer's impression of the amplitude of a sound, which includes the response characteristics of the ear.

natural frequency
See **resonance.**

noise
The terms "noise" and "sound" are often used interchangeably but, generally, *sound* is descriptive of useful communication or pleasant sounds, such as music; whereas *noise* is used to describe dissonance or unwanted sound.

noise reduction coefficient (NRC)
The arithmetical average of the sound absorption coefficients of a material at 250, 500, 1,000 and 2,000 Hz.

octave band
A frequency bandwidth that has an upper band-edge frequency equal to twice its lower band-edge frequency.

peak level
The maximum instantaneous level that occurs over any specified time period.

period (T)

The time (in seconds) required for one cycle of pressure change to take place; hence, it is the reciprocal of the frequency.

pitch

A measure of auditory sensation that depends primarily upon frequency but also upon the pressure and waveform of the sound stimulus.

power

See **sound power.**

pure tone

A sound wave with only one sinusoidal pressure change per unit time.

random-incidence sound field

See **diffuse sound field.**

random noise

A noise made up of many frequency components whose instantaneous amplitudes occur randomly as a function of time.

resonance

The operating condition of an oscillating system that requires the least amount of energy to sustain oscillations. Any change in the driving frequency will cause a decrease in the response of the system.

reverberation

The persistence of sound after the generation of the sound has stopped. The reverberation of a space is specified by the *reverberation time*, which is the time required after the source has stopped radiating sound for the rms sound pressure to decrease 60 dB from its steady state level.

root mean square sound pressure

The root mean square (rms) value of a changing quantity, such as sound pressure, is the square root of the mean of the squares of the instantaneous values of the quantity.

sound

See **noise.**

sound intensity

The average rate at which sound energy is transmitted through a unit area normal to the direction of sound propagation. The units used for sound intensity are joules per square meter per second. Sound intensity is also expressed in terms of a level (sound intensity level, L_I) in decibels referenced to 10^{-12} watts per square meter.

sound power (W)

The total sound energy radiated by a source per unit time. Sound power is normally expressed in terms of a level (sound power level L_W) in decibels referenced to 10^{-12} watts.

sound pressure (p)

The rms value of the pressure changes above and below atmospheric pressure when used to measure steady-state noise. Short-term or impulse-type noises are described by peak pressure values. The units used to describe sound pressures are newtons per square meter (N/m^2), dynes per square centimeter (dyn/cm^2), microbar (μbar), or pascals (Pa). Sound pressure is also described in terms of a level (sound pressure level, L_p) in decibels referenced to 2×10^{-5} N/m^2.

spectrum

A series of energy components of a complex signal arranged in order of frequency (or wavelength).

standing waves

Periodic waves that have a fixed distribution in the propagation medium.

transmission loss (TL)

May be defined as ten times the logarithm (to the base 10) of the ratio of the incident acoustic energy to the acoustic energy transmitted through a barrier.

ultrasonic

A sound with a frequency higher than that of audible sound.

velocity

The time rate of movement of an object or disturbance as it travels away from a reference position in a given direction. Sound velocity (c) varies directly with the square root of the density and inversely with the compressibility of the transmitting medium, the velocity of sound is usually considered constant under normal conditions. For example, the velocity of sound is approximately 343 meters/sec (1,130 ft/sec) in air, 1,500 meters/sec (4,700 ft/sec) in water, 4,000 meters/sec (13,000 ft/sec) in wood and 5,000 meters/sec (16,500 ft/sec) in steel.

volume unit (VU)

A logarithmic (dB) unit used for expressing the magnitude of a complex waveform such as that of speech or music. A specific meter damping is required for measurement.

wavelength (λ)

The distance required to complete one pressure cycle. The wavelength, a very useful tool in noise control, may be calculated from known values of frequency (f) and velocity (c) as $\lambda = c/f$.

white noise

Noise that has an essentially random spectrum with equal energy per unit frequency bandwidth over a specified frequency band.

PHONOLOGICAL TERMS

articulation

The process of producing speech sounds by the movement of structures in the mouth and oral cavity.

articulators

The structures of the mouth and oral cavity that operate to produce speech sounds.

Bernoulli effect

A reduction in pressure that always develops just behind a rapidly moving mass.

breathy speech

Speech sound produced when the vocal folds are adducted anteriorly and narrowly abducted posteriorly resulting in a combination of voicing with friction noise.

cartilage

Material similar to but softer and more flexible than bone.

cognates

Two speech sounds that are produced in the same way and only differ in whether or not normal voicing is used.

cornu

See **horns.**

creak

Type of sound produced when thick, relaxed vocal folds are pulled tightly together and air puffs are released irregularly from the center of the glottis resulting in popping sounds.

cricoid cartilage

A ring of cartilage that is thicker posteriorly and shaped somewhat like a signet ring, which forms the inferior support for the larynx and rests atop the trachea.

cricothyroid muscles

A set of muscles stretched between the cricoid and thyroid cartilages which tend to rock the thyroid cartilage forward and down when they are contracted, thereby lengthening and stretching the vocal folds.

epiglottis

A muscular, doorlike flap arising from the larynx which closes the larynx and laryngopharynx during eating and opens them during respiration.

esophagus

A muscular tube which directs food through the chest cavity and into the stomach for digestion.

glottal stop

Sudden cessation or release of normal voicing by firmly adducted vocal folds.

glottis

The opening between the vocal folds.

horns

Projections of the thyroid cartilage.

inferior horns

Downward, hinge-like projections of the thyroid on which this cartilage can rock forward (anteriorly) and down (inferiorly) depending on the action of various laryngeal muscles.

interarytenoids

Muscles stretched between the left and right arytenoid cartilages which tend to pull these cartilages together adducting the vocal folds and closing the glottis.

jet noise

Sound produced when an airflow is forced quickly through a narrow opening, resulting in a friction noise of fast moving air particles rubbing against slow moving air particles.

larynx

A cartilaginous structure which protects the lungs from food and other foreign particles.

modes of phonation

Various different operating patterns for producing speech sounds.

nasopharynx

A passageway at the posterior inferior portion of the nasal cavity through which air passes in respiration.

normal voicing
Siren-like sound produced by the vocal folds when they are adducted with a moderate amount of vocal fold tension.

oropharynx
A passageway at the posterior end of the mouth cavity and inferior to the nasopharynx, through which air passes during respiration and food passes during eating.

phonation
Producing sound at the larynx for speech production.

sub-glottal space
Superior part of the trachea beneath the glottis.

siren
A device which, when operated, releases a series of air puffs.

superior horns
Upward (superior) projections of the posterior thyroid cartilage that support the hyoid bone.

thyroid cartilage
A large, shield-shaped cartilage that forms the anterior support and protection for the larynx.

trachea
Tube, supported by rings made of cartilage, which serves as an air passageway between the larynx and the bronchi of the lungs.

vocal folds
Muscular tissues within the larynx which can close the larynx to trap food which accidentally got past the epiglottis.

vocal fry
See **creak.**

voiced cognate
Cognate speech sound with normal phonation.

voiced speech
See normal voicing.

voiceless cognate
Cognate speech sound without phonation.

voiceless speech
Sound produced when the vocal folds are completely abducted and air passes freely and soundlessly through the glottis while some con-

figuration of the vocal tract causes noise at a point superior to the larynx.

whispered speech
Speech sound produced by narrowly opening the vocal folds causing a noisy airflow that results in a characteristic hissing sound.

AUDIOLOGICAL TERMS

acoustic reflex
Reflexive contraction of the middle ear muscles in response to a loud sound.

annular ligament
Ligament that seals the stirrup footplate into the oval window allowing the stirrup to move freely without leaking any fluid from the inner ear.

anvil
The middle bone of the ossicular chain.

apex
That end of the cochlea's membranous labyrinth farthest from the middle ear.

apical turn
See **apex.**

area ratio
The area of the eardrum membrane divided by the area of the stirrup footplate.

attic
The superior extension of the middle ear space.

auricle
See **pinna.**

base
That end of the cochlea's membranous labyrinth nearest the middle ear

basal turn
See **base.**

basilar membrane
A sheet of soft tissue that extends from the bony shelf to the outer wall of the cochlea's osseous labyrinth and divides the scala media from the scala tympani.

bony shelf
A ledge of bone that extends from the modiolus into the osseous labyrinth of the cochlea.

bony portion
The inner (medial) third of the external ear canal which is supported by the temporal bone.

buckling effect
The force increase caused by membrane bending.

cartilaginous portion
The outer two-thirds of the external ear canal.

cerumen
A soft wax secreted by the sebaceous glands.

cilia
Small hairlike projections of hair cells.

cochlea
A structure within the temporal bone whose primary function is transducing sound waves into neural signals.

cochlear duct
See **scala media.**

Corti's organ
See **organ of Corti.**

ear canal entrance
An opening extending medially from the auricle.

eardrum (or tympanum)
The eardrum membrane and middle ear space.

eardrum membrane
A cone-shaped membrane that seals the medial end of the external ear canal.

eddy current
Swirling fluid motion in the cochlea that controls the pattern of wave movement.

endolymph

An ionic fluid that fills the scala media except for Corti's organ, has a potential of +80 mV, and is kept separate from the perilymph of the other two scalae.

Eustachian tube

A tube extending from the nasopharynx to the middle ear that aerates the middle ear.

external ear canal

A tube extending medially from the concha that ends in the eardrum membrane.

hair cells

Sensory transducers that convert sound waves into nerve impulses.

hammer

The most lateral of the ossicles.

handle

Long inferior projection of the hammer.

head of hammer

the more massive superior part of the hammer.

head of stirrup

lateral tip of the stirrup.

helicotrema

an opening connecting the scala vestibuli with the scala tympani at the apex of the cochlea.

impedance transformer

Any system that bridges he impedance difference between two media.

incompressible

A property of liquids that cannot be made more dense than they already are.

incus

See **anvil.**

inner

place descriptor for the organ of Corti meaning toward the modioulus.

inner hair cells
thick, bag-shaped sensory cells with cilia that extend beyond the reticular lamina toward the tectorial membrane.

lateral semicircular canal
one of three semicircular canals of the balance mechanism which is in the horizontal plane and forms the promontory in the middle ear space.

lever ratio
The length of hammer divided by the length of anvil.

localization
The ability to determine where a sound is coming from.

long process
Inferior projection of the anvil.

malleus
See **hammer.**

manubrium
See **handle.**

mechanical ear
The outer and middle ears.

middle ear
A cavity in the temporal bone containing the ossicles.

middle ear muscles
Two muscles that work to reduce the amplitude of intense vibrations before they enter cochlea.

modiolus
A column of bone that forms the center post of the cochlea's osseous labyrinth.

pinna
The visible part of the ear.

organ of Corti
A trapezoid-shaped structure where transduction takes place that rests on the basilar membrane of the scala media.

ossicles
Three small bones in the middle ear.

ossicular chain
Term for the three ossicles taken together.

outer ear
That part of the ear including the pinna and the external ear canal.

outer hair cells
Long, tube-shaped sensory cells with cilia that extend beyond the reticular lamina toward the tectorial membrane.

perilymph
Fluid filling both the scala vestibuli and the scala tympani and having a neutral potential (about 0 mV).

posterior semicircular canal
One of three semicircular canals of the balance mechanism which is in the vertical plane and is roughly coronal in orientation.

Reissner's membrane
A sheet of soft tissue that extends from the bony shelf to the outer wall of the cochlea's osseous labyrinth and separates the scala media from the scala vestibuli.

reticular lamina
The surface of Corti's organ nearest the tectorial membrane.

scala media
One of the three ducts of the cochlea's membranous labyrinth that lies between the scala vestibuli and the scala tympani and contains the organ of Corti.

scala tympani
One of the three ducts of the cochlea's membranous labyrinth that begins at the round window in the basal turn and extends to the apex.

scala vestibuli
One of the three ducts of the cochlea's membranous labyrinth that begins at the oval window in the basal turn and extends to the apex.

shearing action
The sliding of the tectorial membrane over the body of the organ of Corti when sound waves enter the cochlea.

shearing force
The force arising from the shearing action within the organ of Corti.

stapes
See **stirrup.**

stirrup

The most medial of the ossicles.

tectorial membrane

A soft structure within the scala media which is hinged at the limbus and forms a roof or covering over the body of the organ of Corti.

transduction

The process of converting energy from one form to another.

tympanum

See **eardrum.**

Appendix D
BIBLIOGRAPHY

This Bibliography is presented as a source of further detail on most of the subjects covered in this text. Many are landmark documents which served to establish a principle or to result in establishment of Standards in the particular field. Each of these documents will also lead to additional references where greater detail on a particular subject may be found.

BOOKS

American Industrial Hygiene Association. *Industrial Noise Manual*, Akron: American Industrial Hygiene Association, 1975.

Beranek, L. L. *Acoustic Measurements*, New York: John Wiley and Sons, Inc., 1949.

Beranek, L. L. *Music, Acoustics and Architecture*, New York: John Wiley and Sons, Inc., 1962.

Employer's Insurance of Wausau. *Guide for Industrial Audiometric Technicians*, Wausau, WI: Employer's Insurance of Wausau.

Glorig, A. *Noise and Your Ear*, New York: Grune and Stratton, 1958.

Harris, C. M., ed. *Handbook of Noise Control*, New York: McGraw-Hill, 1957.

Kinsler, L. E., A. R. Frey, A. B. Coppens, and J. V. Sanders. *Fundamentals of Acoustics*, New York: John Wiley and Sons, 1982.

Kryter, K. D. *The Effects of Noise on Man*, New York: Academic Press, 1970.

Parkin, P. H. et al. *Acoustics, Noise and Buildings*, London: Faber and Faber, Inc., 1979.

Rossing, T. D. *The Science of Sound*, Reading: Addison-Wesley Publishing Co., 1990.

Satatoff, J. and P. Michael. *Hearing Conservation*, pp. 181–186. Springfield: Charles C. Thomas, 1973.

STANDARDS

Through mutual agreement between instrument manufacturers and research laboratories, standard specifications and procedures are established. This list includes some of the most frequently used documents. Many others are available from the Standards Secretariat of the Acoustical Society of America.

American National Standard Criteria for Permissible Ambient Noise during Audiometric Testing ANSI S3.1-1977 American National Standards Institute, New York

American National Standard Guide for the Evaluation of Human Exposure to Whole-Body Vibration ANSI S3.18-1979 American National Standards Institute, New York

American National Standard Method for the Measurement of Real-Ear Protection of Hearing Protectors and Physical Attenuation of Earmuffs ANSI S3.19-1974 (R1979) (See also ANSI S12.6-1984) American National Standards Institute, New York

American National Standard Methods for the Calculation of the Articulation Index ANSI S3.5-1969 (R1978) American National Standards Institute, New York

American National Standard Methods for the Measurement of Sound Pressure Levels ANSI S1.13-1971 (R1976) American National Standards Institute, New York

American National Standard Preferred Frequencies, Frequency Levels, and Band Numbers for Acoustical Measurements ANSI S1.6-1984 American National Standards Institute, New York

American National Standard Specification for Sound Level Meters ANSI S1.4-1983 American National Standards Institute, New York

American National Standard Specifications for Audiometers ANSI S3.6-1969 (1973) American National Standards Institute, New York

American National Standard Specifications for Octave, Half-Octave and Third-Octave Band Filter Sets ANSI S1.11 (R1976) American National Standards Institute, New York

International Organization for Standardization, *ISO Recommendation R 1996, Acoustics*, "Assessment of Noise with Respect to Community Response," 1st Edition, ISO/R 1996–1971 (E), May 1971.

ARTICLES, PROCEEDINGS, AND REPORTS

The following articles are primarily related to environmental noise. Many are reports of original research and are landmark documents that led to the establishment of standards and/or legislation on noise exposure.

American Speech and Hearing Association. "Noise as a Public Health Hazard." *American Speech and Hearing Association* 4 (February 1969):105–109.

Bell, A. "Noise—An Occupational Hazard and Public Nuisance." *Public Health Paper*, no. 30, Geneva, Switzerland (1966).

Bonvallet, G. L. "Levels and spectra of Traffic, Industrial and Residential Area Noise." *Journal Acoustical Society America* 23 (July 1951):435–439.

Bottom, C. G. and D. M. Waters. "A Social Survey Into Annoyance Caused by the Interaction of Aircraft Noise and Traffic Noise." *TT7102 Department of Transport Technology, Loughborough University of Technology, England.*

Carpenter, A. "How Does Noise Affect the Individual?" *Impulse* no. 24 (1964).

Cohen, A. "Effects of Noise on Performance." *Proceedings of the International Congress of Occupational Health, Vienna, Austria* (1966):157–160.

Donley, R. "Community Noise Regulation." *Sound and Vibration* (February 1969).

Eagles, E. L. and L. G. Doerfler. "Hearing in Children: Acoustic Environment and Audiometer Performance." *Journal Speech Hearing Research* 4 (1961):149.

Felton, J. S. and C. Spencer. "Morale of Workers Exposed to High Levels of Occupational Noise." *American Industrial Hygiene Association Journal* 22 (1961):136–147.

Galloway, W. J. and A. C. Pietrasanta. "Land Use Planning Relative to Aircraft Noise." *Technical Report no. 821*, Bolt, Beranek and Newman, Inc. (October 1964). [Also published by the Department of Defense as AFM 86-5, TM 5-365, NAVDOCKS P-98.]

Grandjean, E. "Physiologische and Psychophysiologische Wirkingen des Larms." *Menschen Umwelt* no. 4 (1960).

Intersociety Committee on Industrial Audiometric Technician Training, "Guide for Training of Industrial Audiometric Technicians." *American Industrial Hygiene Association Journal* 27 (1966):303.

Knight, J. J. "Normal Hearing Threshold Determined by Manual and Self-Recording Techniques." *Journal Acoustical Society America* 39 (1966):1184.

Kryter, K. D. "Speech Communication in Noise." AFCRCOTRO54-52 (Air Force Cambridge Research Center, Air Research and Development Command, Washington, DC), (1955).

Kryter, K. D. "Psychological Reactions to Aircraft Noise." *Science 151* 3716 (1966):1346–1355.

Lawther, A. and D. W. Robinson. "An Investigation of Tests for Susceptibility to Noise-Induced Hearing Loss." *Institute of Sound and Vibration Research ISVR Technical Report 138* University of Southhampton (1986).

Lawther, A. and D. W. Robinson. "Further Investigation of Tests for Susceptibility to Noise-Induced Hearing Loss." *Institute of Sound and Vibration Research ISVR Technical Report 149* (1987).

Lawther, A. and D. W. Robinson. "Potential Factors in Resistance to Noise-Induced Hearing Loss." *Proc. Institute of Acoustics*, 9 (3), 217–219 (1987).

Meeker, W. F. "Active Ear Defender Systems: Component Considerations and Theory." *Wright Air Development Center TR 57-368* vol. 1 (1958).

Meeker, W. F. "Active Ear Defender Systems: Development of a Laboratory Model." *Wright Air Development Center TR 57-368* vol. 2 (1959).

Michael, P. L. "Ear Protectors—Their Usefulness and Limitations." *Arch. Environ. Health* 10 (1965):612–618.

Michael, P. L. "Standardization of Normal Hearing Thresholds." *Journal Occupational Medicine* 10 (1968):67.

Pollack, I. and J. M. Pickett. "Masking of Speech by Noise at High Sound Levels." *Journal Acoustical Society America* 30 (1958):127–130.

Riley, E. C., J. H. Sterner, D. W. Fassett and W. L. Sutton. "Ten Years Experience with Industrial Audiometry." *American Industrial Hygiene Association Journal* 22 (1961):151.

Robinson, D. W. "Towards a Unified System of Noise Assessment." *Journal Sound and Vibration* 14 (3) (1971):279–298.

Rosen, S., D. Plester, A. El-Mofty and H. V. Rosen. "High Frequency Audiometry in Presbycusis: A Comparative Study of the Maaban Tribe in the Sudan with Urban Populations." *Arch. Otolaryngol.* 79 (1964):1.

Scott, H. H. "The Degenerative Sound Analyzer." *Journal Acoustical Society America* 11 (1939):225.

Shetalov, N. N., A. D. Sartausv and K. V. Glotov. "On the State of the Cardiovascular System under Conditions of Exposure to Continuous Noise." *Labor Hygiene Occupational Diseases* 6 (1962):10–14.

Simpson, M. and D. Bishop. "Community Noise Measurements in Los Angeles, Detroit and Boston." *Bolt, Beranek and Newman Report No. 2078* June 1971.

Tracor Laboratories Staff. "Community Reaction to Airport Noise." *Final Report, Vol. 1, Tracor Document no. T-70-AU-7454-U*. September 1970.

von Gierke, H. E. and D. R. Warren. "Protection of the Ear from Noise: Limiting Factors." *Benox Report*, The University of Chicago, Chicago, IL, December 1953.

Webster, J. C. "Effects of Noise on Speech Intelligibility." *American Speech and Hearing Association Reports* 4 (1969): 49–73.

Webster, J. C. "SIL—Past, Present and Future." *Sound and Vibration* (August 1969).

Webster, J. C. "Updating and Interpreting the Speech Interference Level (SIL)." *J. Audio Engr. Soc.* 18 (1970):114–118, (April 1970).

Whittle, I. S. and M. E. Delaney. "Equivalent Threshold Sound-Pressure Levels for the TDH39/MX41-AR Earphone." *Journal Acoustical Society America* 38 (1966):1187.

Wyle Laboratories Research Staff, "An Evaluation of Methods for Scaling Aircraft Noise Perception." *Wyle Research Laboratories Report WR70-17*. May 1971.

Young, R. W. and A. Peterson. "On Estimating Noisiness of Aircraft Sounds." *Journal Acoustical Society America* 45 (1969):834–838.

GOVERNMENT DOCUMENTS

These reports include noise studies performed by Government agencies and official regulation established as a result of these studies and other noise research.

American Council of Governmental Industrial Hygienists, "Threshold Limit Values and Biological Exposure Indicies for 1988–89," (ACGIH), 1988.

Borsky, P. N. *Community Reactions to Air Force Noise*, WADC Technical Report 60-689, Parts 1 and 2, Wright Air Development Depot, Wright-Patterson Air Force Base, March 1961.

Committee on the Problem of Noise "Social Survey in the Vicinity of London (Heathrow) Airport," *Noise*, Appendix XI, presented to Parliament by the Lord President of the Council and Minister for Science, Her Majesty's Stationery Office, July 1963 rev. 1966.

DOT Office of Noise Abatement: *A Study of the Magnitude of Transportation Noise Generation and Potential Abatement.* vol. 3, Report OST-ONA-71-1, Washington, November 1970.

Federal Aviation Agency: *Noise Standards: Aircraft Type Certification.* Federal Aviation Regulations, Part 36, Washington, November 1969.

Michael, P. L., and D. F. Bolka, The Pennsylvania State University, Environmental Acoustics Laboratory. *An Objective Method for Evaluating Ear Protectors.* Final Report to Department of Health, Education and Welfare, U.S. Public Health Service, June 15, 1972

Mills, C. H. G. and D. W. Robinson. "The Subjective Rating of Motor Vehicle Noise," *Noise*, Appendix IX, presented to Parliament by the Lord President of the Council and Minister for Science by Committee on the Problem of Noise, Her Majesty's Stationery Office, July 1963.

Outline for Course Leading to Accreditation as an Occupational Hearing Conservationist. Council for Accreditation in Occupational Hearing Conservation, Haddon Heights, NJ.

Robinson, D. W. *The Concept of Noise Pollution Level*, NPL Aero Report Ac 38, National Physical Laboratory, Aerodynamics Division, March 1969.

Rosenblith, W. A., K. N. Stevens, et al. *Handbook of Acoustic Noise Control,* vol. 2, *Noise and Man,* WADC TR-52-204, Wright Air Development Center, Wright-Patterson Air Force Base, 1953.

Stevens, K. N., A. C. Pietrasanta, et al. *Procedures for Estimating Noise Exposure and Resulting Community Reactions from Air Base Operations,* WADC TN-57-10, Wright Air Development Center, Wright-Patterson Air Force Base, 1957.

U.S. Congress. *Occupational Safety and Health Act of 1970.* Public Law 91-596, 91st Cong., Washington, 29 December 1970.

U.S. Congress. *Walsh-Healey Public Contracts Act.* Federal Register, Vol. 34, No. 96, Washington, 20 May 1969.

U.S. Department of Health, Education and Welfare, National Institute for Occupational Safety and Health. *Criteria for a Recommended Standard Occupational Exposure to Noise.* NIOSH 73-11001, Washington, 1972.

U.S. Department of Labor. *Noise Control—A Guide for Workers and Employers,* Occupational Safety and Health Administration, Office of Information, no. OSHA-3048, 1980.

U.S. Department of Labor. *Occupational Nose Exposure: Hearing Conservation Amendment.* Occupational Safety and Health Administration, 29 CFR Part 1910. Federal Register, *46,* no. 162, August 21, 1981.

U.S. Department of Labor. *Occupational Noise Exposure: Hearing Conservation Amendment.* Occupational Safety and Health Administration, 29 CFR Part 1910. Federal Register, *48,* no. 46, March 3, 1983.

U.S. Department of Labor. *Occupational Safety and Health Standards,* Williams-Steiger Occupational Safety and Health Act of 1970, Federal Register 36, p. 10518, 1971.

U.S. Department of Labor. *Safety and Health Standards for Federal Supply Contracts,* Walsh-Healey Public Contracts Act, Federal Register, 34 p. 7948, 1969.

U.S. Department of Labor. Williams-Steiger Occupational Safety and Health Act of 1970: *Occupational Safety and Health Standards.* Federal Register. Washington, 29 May 1971.

U.S. Department of Transportation, Office of Noise Abatement. *A Study of the Magnitude of Transportation Noise Generation and Potential Abatement,* vols. 1 thru 8, Department of Transportation, Washington, November 1970.

U.S. Department of Transportation, Office of Noise Abatement. *Noise and Vibration Characteristics of High Speed Transit Vehicles*, Technical Report OST-ONA-71-7, Department of Transportation, Washington, June 1971.

U.S. Environmental Protection Agency. *Community Noise.* NTID 300.3, Washington, 31 December 1971.

U.S. Environmental Protection Agency. *Community Noise Fundementals-Independent Study by Correspondence*, U.S. Environmental Protection Agency, Office of Noise Abatement and Control. Final Report on Contract No. 68-01-3895. (The Environmental Acoustics Laboratory, The Pennsylvania State University), 1977, 160 Pp.

U.S. Environmental Protection Agency. *Information on Levels of Environmental Noise Requisite to Protect Public Health and Welfare with and Adequate Margin of Safety.* EPA 550/9-74-004, Washington, March 1974.

U.S. Environmental Protection Agency: *Laws and Regulatory Schemes for Noise Abatement.* NTID 300.4, Washington, 31 December 1971.

U.S. Environmental Protection Agency. *Model Community Noise Control Ordinance.* EPA Document 550/9-76-003, Washington, September 1975.

U.S. Environmental Protection Agency. *Noise from Transportation Systems, Recreation Vehicles and Devices Powered by Small Internal Combustion Engines*, WR71-17, Wyle Laboratories, Office of Noise Abatement and Control, U.S. Environmental Protection Agency, Washington, November 1971.

U.S. Environmental Protection Agency. *Report to the President and Congress on Noise.* EPA Document 92: 1–46 and 63: 1–48, Washington, February 1972.

INDEX